Tea Tree Oil for Health & Well-Being

Susanne Poth

Tea Tree Oil for Health & Well-Being

Sterling Publishing Co., Inc.
New York

Library of Congress Cataloging-in-Publication Data Available

Photo credits: Archiv für Kunst und Geschichte (AKG), Berlin, p. 11; **Beat Ernst,** Basel: pp. 9, 20; **Bergland Pharma,** Heimertingen, p. 13: **Bildarchiv Huber,** Garmisch-Partenkirchen, p. 45 (Picture Bank); **IFA-Bilderteam**, München: pp. 1 (F. Prenzel), 2 (Gottschalk), 7, 34 (Russia), 24 (Bumann), 33, 70 (Diaf), 37, 40 (IPP), 48 (Behr), 72 (Marc); **Ulrich Niehoff,** Bienenbüttel, pp. 49, 51, 52, 54, 73, 75, 80, 85; **Okapia**, Frankfurt: pp. 6, 22 (H. Reinhard), 87 (Emu): **Dr. L. Reinbacher,** Kempten: p.27; **Silvestris GmbH,** Kastl/Obb: p. 67 (Lindenberger); **Tebasan,** Norderstedt bei Hamburg: p. 6, 15, 17: **Thursday Plantation,** München: p. 14; **FALKEN Archiv**, Niedernhausen: pp. 64(U. Zöltsch), 76 (C. Steiner), 78, 83 (Kienitz & Grabis).

10 9 8 7 6 5 4 3 2 1

Published by Sterling Publishing Company, Inc.
387 Park Avenue South, New York, N.Y. 10016
Originally published in Germnany under the title *Gesundheit und Wohlbefinden durch Teebaumöl* and © 1997 by FALKEN Verlag, 65527 Niedernhausen/Ts.
English translation © 1999 by Sterling Publishing Company, Inc.
Distributed in Canada by Sterling Publishing
c/o Canadian Manda Group, One Atlantic Avenue, Suite 105
Toronto, Ontario, Canada M6K 3E7
Distributed in Great Britain and Europe by Cassell PLC
Wellington House, 125 Strand, London WC2R 0BB, England
Distributed in Australia by Capricorn Link (Australia) Pty Ltd.
P.O. Box 6651, Baulkham Hills, Business Centre, NSW 2153, Australia
Printed in Hong Kong

Sterling ISBN 0-8069-4848-5

Tea Tree Oil for Health & Well-Being

Contents

Preface 8

A Remedy from Down Under 9

The Tea Tree Oil Success Story 10

First Aid for Australia's Aborigines 10

Tea Tree Imparts a Good Taste to Captain Cook's Beer 10

Western Science Discovers Tea Tree Oil 11

The Comeback of the "Miracle of the Island Continent" 12

The Future of Tea Tree Oil 12

Tea Tree from a Botanical Point of View 13

The Weed from the Swamps 13

Tea Tree Plantations 14

The Essential Oil of the Tea Tree 16

Extracting the Oil by Distillation 16

Essential Oils in General 16

The Components of Tea Tree Oil 17

Two Components Form the Quality Standard 18

Tea Tree and Its Relatives 19

Cajeput Oil 20

Niaouli Oil 21

Manuka Oil 22

Ravensara Oil 23

Tea Tree Oil Research 24

Laboratory Tests 25

No Chance for Bacteria 25

Antifungal Effects 27

Can a Virus Be Stopped, Too? 27

Additional Useful Properties 28

Clinical Studies 29

The Ideal Antiseptic? 29

Healing Boils 30

An Aid for Vaginal Infections 30

Relief for Foot Problems 31

Therapeutic Value with Acne 32

How Does Tea Tree Oil Work? 34

A Comparison with Antibiotics 35

Tolerance of This Essential Oil 36

External Application Is Safe 36

Not Recommended for Internal Use 38

Overall Safety 39

Self-Treatment with Tea Tree Oil 40

Oil, Hydrolates, and Solutions 41

Areas of Use from A to Z 42

Abscesses 43
Acne 43
Athlete's Foot 47
Burns, Minor 49
Chickenpox and Shingles 50
Colds 51
Colds—Prevention 51
Coughs 52
Cuts and Abrasions 53
Dandruff 54
Fungal Infections of the Vagina 55
Fungus of the Nails 59
Gangrene 61
Inflammation of the Gums 61
Inflammation of the Nail Wall 62
Inflammation of the Skin 62
Insect Bites 63
Insect Repellent 63
Intestinal Fungi 65
Lip Blisters 66
Neurodermatitis 67
Pains in the Muscles and Joints 70
Rosacea 71
Sore Throat and Hoarseness 71
Toothache 72
Warts 72

Other Possible Uses 73

First-Aid Travel Kit 73
Aromatherapy 74
Animal Care 76
Cleaning and Disinfecting in the Home 76

Applying Tea Tree Oil from Head to Toe 78

Tea Tree Oil in Cosmetic Products 79

Cleansing and Care of the Skin 79
For Beautiful Hair and a Healthy Scalp 80
Prevention of Body Odor 81
Foot Care 81
Skin Care After Sunbathing 84
Dental Hygiene 84

APPENDIX 87

Glossary 88

Checklist: Staying Healthy When Traveling 90

Index 92

Preface

"The greatest hindrance to the progress of science is the desire to see the results of success immediately."

Georg Christoph Lichtenberg

What does this wise statement, made by a physicist and a philosopher, have to do with tea tree oil, which people have been using for thousands of years? Until recently, tea tree oil was used to treat a limited number of ailments—mainly bacterial and, to a certain extent, fungal disorders—and just a few tons of the oil were sold each year. But in the last few years, business worth millions of dollars has resulted from the boom in tea tree oil, which also seems to have inspired numerous authors. Research has shown that tea tree oil can have positive as well negative effects, yet most of these authors have been quick to praise the positive results, while simply ignoring any negative consequences.

Countless new possibilities for using tea tree oil have been recommended lately, although very few of them have ever been tested and investigations into results have been glossed over or even falsified. Hence, as of late, tea tree oil has joined the ranks of the world's new magic potions.

But as could be expected, voices raised in contradiction were soon to be heard. The danger of allergic reactions engendered criticism of tea tree oil, and hasty, overconfident claims made about the oil damaged its reputation among many scientists, physicians, pharmacists, and nonmedical practitioners.

The truth is that tea tree oil is not a magic potion, and to date there hasn't been sufficient research into its healing properties. Nonetheless, the research results we do have indicate that this essential oil is a valuable addition to medicinal plant therapy.

At this point, I would like to extend my special thanks to Professor Jürgen Reichling, who provided scientific information and photos for the book as well as the most recent research findings on tea tree oil. Furthermore, I would like to thank Inge Andres for her contribution concerning related tea tree oils and for a most informative exchange of ideas on the subject. I would also like to thank pharmacist Anette Schenk for her continuous support.

Pharmacist Susanne Poth

A Remedy from Down Under

The Tea Tree Oil Success Story

Many essential oils are outstanding in combating bacterial and fungal infections, and this is not just a recent discovery. Essential oils from plants are well known for their disinfectant properties, and have been used for thousands of years in treating cuts and abrasions, animal bites, and insect bites and stings. Which of these natural healers people used depended on the region where they lived; for example, lemon grass oil was first used in Asia, and Europeans used thyme or chamomile. The tea tree oil success story starts in Australia with the original inhabitants of the country, the Aborigines.

First Aid for Australia's Aborigines

The Bundjalung Aborigines, who lived as hunter-gatherers in the north of New South Wales (the southeast part of Australia), were probably the first to discover the therapeutic properties of tea tree oil. Their special way of preparing this medicinal plant has been handed down for generations right up to modern times: the leaves of the tea tree are crushed and ground in order to break open glandular dots in them and thus to release the essential oil they contain; then the leaves are soaked in water and used in compresses or infusions. The Aborigines have always used this method to disinfect wounds and animal and insect bites as well as to bring relief to sore throats.

The Australian Aborigines have used tea tree traditionally as a medicinal plant.

The Aborigines also know about inhaling the essential oil. They do this by either breathing in the steam from the crushed leaves or inhaling the smoke from tea tree twigs burning in a small fire. Both methods are effective remedies for colds. To produce ointments and pastes, they mix tea tree leaves with animal fat or clay. Then they are rubbed onto the body or used in compresses to treat a variety of ailments.

Tea Tree Imparts a Good Taste to Captain Cook's Beer

Europeans became aware of tea tree during their first expeditions to Australia. Captain James Cook of the British Royal Navy, who sailed around the world and landed on the southeast coast of Australia in 1770, encountered thick groves of trees having aromatic leaves, which he described in his diary of the trip as "tea plants." The name "tea tree" was first coined by his crew, who brewed an aromatic tea from its leaves. Sir Joseph Banks, a botanist with the expedition, took some of the leaves back to England, where he presented them to the Western world for the first time. It's been said that during his second expedition, Captain Cook brewed beer from spruce leaves, which turned out to be very bitter, so he improved on the taste of the beer with the aid of tea tree leaves. However, the medicinal properties of tea tree were not recognized at that time, although the seamen must have seen the Aborigines using it for healing.

Later, the white settlers began to emulate the Aborigines in their use of this indigenous bush medicine, treating many complaints, including cuts, lice and fleas, insect bites, and infections, with the oil.

The name "tea tree" dates back to the expeditions of English navigator and explorer James Cook (1728–1779).

Western Science Discovers Tea Tree Oil

It was only in the 1920s that this traditional use of tea tree oil was recognized by Western science. Dr. Arthur R. Penfold, an Australian chemist, together with his team of researchers, distilled the essential oil from the leaves of the tea tree, and then analyzed its effect on fungi and bacteria cultures in the laboratory. They also made a comparison of different tea tree oils. The results of their research were sensational: they found that the effect of this essential oil on bacteria exceeded by far the antiseptic potential of the disinfectant carbolic acid (today, known under the name of "phenol"), which was in wide use at the time.

The results of this research, published by Penfold in 1925 and 1929, were the reason that this medicine from the outback finally began to be taken seriously in the Western world. The publications triggered a wave of further tests not only in Australia but also in England, France, and America. Tea tree oil became increasingly popular with doctors, dentists, and veterinarians, because it is well tolerated by the body and is a very effective disinfectant, and it soon became known as "Australian Gold."

However, the interest in this "cure-all" declined rapidly after World War II. This was the beginning of chemical disinfectants, the sulfonamides, and antibiotics. Phytotherapy (plant medication) was regarded as "old wives' tales," and tea tree oil research projects came to a halt. Furthermore, it didn't help that there was a great deal of variety in the composition and quality of the natural oils that were distilled from the leaves of the different kinds of tea trees growing in the wild, because the distilling took place under very simple conditions that were difficult to monitor accurately.

The Comeback of the "Miracle from the Island Continent"

When research into natural healing resumed in the 1960s in America, the demand for tea tree oil was practically nil. At that time, the first clinical tests with tea tree oil were conducted on boils and carbuncles, infections of the vagina, and (10 years later) disorders of the feet. The increase in both the side effects of, and resistance to, chemically based medicines, together with a growing interest in holistic medicine, formed the basis for the rediscovery of traditional medicines and healing methods.

The trend toward natural healing has revived an interest in tea tree oil.

Nowadays, efforts are being made to test the effects of tea tree oil by means of modern, well-monitored research methods, and norms are being established for the quality of the oil, thus diminishing its reputation as a "primitive medicine." In the meantime, the "Miracle of the Island Continent," as tea tree oil has been called by some clever promoters, has become a multimillion-dollar business and is one of Australia's most lucrative exports. Not only can you buy the pure essential oil itself, but creams, gels, ointments, shampoos, body lotions, foot balm, and toothpaste containing tea tree oil are available; in some countries, even vaginal suppositories with tea tree oil are on the market. Future plans include using large quantities of the oil in the cleaning and disinfecting of air-conditioning systems, to cite one example.

Only 8 tons of tea tree oil were produced in 1940, but this amount increased tenfold by the 1980s, and the most recent figures show an annual production of nearly 1,000 tons. We can only hope that part of the profits will go back into research, so that we can get some answers to the many questions that remain regarding the "Miracle of the Island Continent."

The Future of Tea Tree Oil

Although tea tree oil is not a miracle drug, it is a natural remedy with many uses, which, if applied correctly, will soon be firmly established in homeopathy. Until that time, however, a great deal of research is needed.

Now that this essential oil has become so popular, it should be even more crucial to inform the consumer about possible side effects instead of continually looking for new areas of use. One cannot stress the medicinal character of this "Australian Gold" while encouraging the consumer to use it every day. It doesn't come as a surprise to discover that an overuse of the oil could cause contact allergies among some people.

In connection with the possible

effects of tea tree oil, the question concerning the right dosages crops up. Aromatherapists generally use a very low dosage: often just a few drops. This is certainly a reasonable approach to take considering the danger of sensitization to the oil; however, it is doubtful whether such a small dosage will always have the desired effect.

In some cases, a higher dosage is necessary if treatment with tea tree oil is to be successful. This was shown in a study made by Australian research scientists (see page 32). For a safe and effective application of tea tree oil, much research still has to be done. This traditional medicine may have a great future, but only after the important questions concerning its side effects and dosages have been addressed.

Tea Tree from a Botanical Point of View

Traditional tea tree oil is an essential oil that is extracted mainly from the leaves of the Australian tea tree, or *Melaleuca alternifolia* (the botanical name). Sometimes it is mixed with the oil of the highly productive *Melaleuca linariifolia* and *Melaleuca dissitiflora*. Melaleuca is one member of the extensive myrtle family, the Myrtaceae, a great number of which are found in Australia. Many of the myrtle plants have long been known to us as spices or medicinal plants—for example, cloves, allspice, and especially eucalyptus, which is well established in many places for the treatment of colds. Thus, tea tree may be said to be predestined to be used as a medicinal plant as a result of its botanical heritage.

The Weed from the Swamps

The swamps in New South Wales in southeast Australia are the original home of the tea tree. *Melaleuca alternifolia* are found only in Australia, mainly in a subtropical climate zone located between the Clarence and Richmond rivers.

Wild tea trees growing in an Australian swamp

Formerly, tea trees were a nuisance to farmers who tried to drain the swamps in order to gain new pastureland for their livestock. For them, the tea tree was a stubborn weed that they couldn't get rid of even when they felled the little trees. They had to completely dig out the deep and widely extended roots of the "weed" so that it would die. If just a bit of the root was left, it took only a few weeks for new shoots to appear, which then grew into a 3- to 4-foot bush or bushy tree within a year; fully grown plants reach heights of between 10 and 20 feet.

The trunk of the tea tree has a paperlike bark, which is why it is also called the paperbark tree. Underneath the white bark, the black trunk is visible in some places—hence, the name *melaleuca*, which is the Greek word for black-and-white.

The thin twigs of the tea tree have narrow, light-green leaves about 2 inches long that are needlelike in appearance and alternate on each side of the twig. The blossoms look like small ears of grain, and they produce capsules containing tiny seeds that are so light in weight that you would have to put about 1,134,000 of them on a pair of scales in order to reach a single ounce. New plants are grown from seed. The essential oil is located mainly in the leaves in tiny, glandular dots, which burst when they are crushed and release the oil that then rapidly loses its strong odor. Small amounts of the essential oil can also be found in the bark of the plant.

The small tea trees are cut down to mere stumps with an automatic harvester.

Tea Tree Plantations

Until long after the end of World War II, when the production of tea tree oil amounted to less than 10 tons per year, the oil was taken only from plants growing wild. The cutters had to work on foot in the swamps, removing the leaves from the branches with a sharp machete. Even so, a single cutter could collect up to a ton of leaves per day. By the end of the 1970s, when the interest in tea tree oil had revived, a successful attempt was made to cultivate the swamp plant on plantations. The tea tree grows very well if it is watered, and the plants continuously yield greater amounts of oil when the seeds are selected carefully.

Today, tea tree plantations can be found along nearly the entire east coast of Australia as well as in other countries, such as South Africa, India, and Malaysia, where farmers have started trying to cultivate the plant.

It takes only 15 months for a small tree on a plantation to be ready for the

Today, tea tree is cultivated on huge plantations.

first harvest: from November onward, the leaf glands are filled to bursting with the essential oil and promise high yields. Essential oil in about 1 to 2 percent of the total leaf weight can be pressed from the fresh young leaves. Toward the beginning of the Australian winter in June, this amount decreases.

Soon after that, new shoots appear. It has been observed that plants that were harvested regularly produced more shoots then uncut ones. But even if this extreme cutting back of the trees doesn't apparently harm them, the nutritive elements in the soil will certainly be leached out at some point in time by this monoculture farming. Another questionable procedure is spraying trees with the insecticide lindane, which is done at many plantations. Just like DDT, it doesn't decompose but accumulates in the food chain, and is therefore a very problematic environmental poison.

The Essential Oil of the Tea Tree

Extracting the Oil by Distillation

Steam is used to isolate the essential oil from the leaves.

Immediately after harvesting, the twigs and leaves are taken to the distillation plant, where they are heated in boilers. The steam penetrates the leaves, absorbs the essential oil, and rises into a system of tubes enclosed in a cooling tower, where it condenses. During this process, the essential oil separates and floats to the top, where it can easily be skimmed off. Afterward, it is cleaned, filtered, and bottled. In this way, about 10 quarts of the light-yellow tea tree oil are extracted from a ton of leaves. The water, which is called "hydrolate," is used, too. It contains a bit of the essential oil as well as those water-soluble substances of the tea tree that were picked up by the steam during the distillation process. The organic material that is left can be spread over the fields as a natural fertilizer.

Essential Oils in General

It's important not to confuse essential oils with fatty oils (such as sunflower oil) or mineral oils (such as paraffin), because all they have in common is their viscous consistency.

The fatty oils—like the essential oils—are nearly impossible to mix with water. However, their consistency as well as their functions are completely different.

Fatty oils are combinations of glycerin molecules with three fatty acids, each of different length. Mineral oils, however, are long chains of hydrocarbons. This means that both oils consist of large, relatively heavy molecules, which do not evaporate and would leave greasy marks on a piece of paper.

Essential oils have many different components, such as terpenes and sesquiterpenes, and their phenols, oxides, and ethers. These small molecules do not have a strong force of attraction toward one another, so they vaporize easily and fill the room with an aroma. Thus, essential oils are also known as ethereal oils, which means "heavenly." The reason for the many different aromas can be found in the countless possibilities of mixing an essence using any number of components.

The importance of the oil for the plants themselves also differs. Fatty oils contain a great deal of energy, which is well known to everyone who counts calories. They are a source of energy for the plant and are stored mainly in the seeds. Essential oils, however, are multifunctional: on the one hand, they lure those insects that transport the plant's seeds, and, on the other hand, they provide excellent protection against insect pests and parasites as well as microbial infestation.

The Components of Tea Tree Oil

An encounter with tea tree oil is very rarely "love at first sight." Its odor, which you have to get used to, turns many people off. However, it gives the impression of being a good medicine: tea tree oil simply smells healthy. The reason for the characteristic smell is to be found in the more than 100 different components that make up this "cocktail of odors." Biologists were aware of some of the components already from other essential oils, such as the 1.8 percent cineole in eucalyptus oil. But other components have been found to date only in tea tree oil.

The quantity of the various components depends largely on the time of harvesting, the soil, the climate, and the distillation procedure.

Tea tree twigs with their typically small, needlelike leaves

Components of Tea Tree Oil (Contents in Percent)			
Alpha-phellandrene	0.44	Globulol	0.86
Alpha-pinene	2.46	Limonene	1.01
Alpha-terpinene	9.56	Myrcene	0.86
Alpha-terpineol	3.01	Para-cymene	2.80
Alpha-thujene	0.83	Sabinene	0.45
Beta-pinene	0.66	Terpinen-4-ol	37.93
1.8-cineole	3.87	Terpinolene	3.45
Delta-3-carenne	1.49	Viridiflorene	1.68
Gamma-terpinene	20.20	Viridiflorol	0.33

Two Components Form the Quality Standard

As you can see from the overview of the components, "Australian Gold" is a complicated cocktail of active substances. Conspicuous is the large percentage of terpinen-4-ol, which has proved to be very effective in combating the bacteria that cause diseases. That's why it is not surprising to discover that when tested, tea tree oils with a high terpinen-4-ol content proved to have a greater antibacterial effect than those with a low terpinen-4-ol content. However, other components, such as linalool and alpha-terpineol, are also effective in the fight against bacteria. In laboratory tests, both components proved successful as antibacterial agents.

Other tests dealing with the relationship of terpinen-4-ol and 1.8-cineole showed that often the two components occur inversely, meaning that a greater amount of terpinen-4-ol indicates that less 1.8-cineole is to be found. Incidentally, the latter is found in large quantities in eucalyptus oil and is therefore also known as "eucalyptol."

The amount of terpinen-4-ol and 1.8-cineole are important quality criteria.

A small amount of cineole together with a large amount of terpinen-4-ol soon became synonymous for quality, because it was the cineole that was held responsible for the skin irritations occasionally caused by tea tree oil.

The Australian publication on medication, the *Australian Standard*, referred to in many books on medication in other countries, states that the maximum amount of cineole should not exceed 15 percent. As a result of selective cultivation, essential oils with less and less 1.8-cineole content and at the same time a

Levels of Quality of Tea Tree Oil Produced on Plantations

1. Products from inspected biological cultivation: Recommended for medicinal purposes (terpinen-4-ol > 35%, 1.8-cineole < 5%).
2. Pharmaceutical quality: Also for medicinal purposes (terpinen-4-ol >35%, 1.8-cineole < 5%); as with levels 3, 4, and 5, traces of insecticides cannot be ruled out.
3. Cosmetic quality: Recommended for skin-care programs (terpinen-4-ol >35%, 1.8-cineole < 5%).
4. Technical quality: For use with animal-care products as well as for cleansing and disinfecting (terpinen-4-ol >30%, 1.8-cineole <10%).
5. Industrial quality: Suitable for disinfecting and cleaning and for biological insecticides (terpinen-4-ol >30%, 1.8-cineole <15%).

These distinctions are based upon criteria established by the Australian Tea Tree Industry Association (ATTIA), which maintains that the terpinen-4-ol content and a low amount of cineole are the most important standards.

higher content of terpinen-4-ol are now being produced.

Farmers, exporters, and buyers who have joined to form the Australian Tea Tree Industry Association (ATTIA) have as an aim the establishment of strict standards for the quality of tea tree oil (see the table on page 18), which is being marketed increasingly in dilution with cheaper oils, or has been polluted by insecticides. They use *Melaleuca alternifolia* only in the production of tea tree oil, although, according to present Australian standards, *Melaleuca liniariifolia* and *Melaleuca dissitiflora* are also permitted. In the meantime, however, it is thought that the harmfulness of cineole was overemphasized and that other components caused skin irritations. According to the latest findings, the 1.8-cineole hysteria was exaggerated, and the resulting one-sided cultivation of plants having a low-cineole content was premature (see page 36, as well).

Tea Tree and Its Relatives

Even though the name suggests it, tea tree is not in any way related to what we usually call tea. Black tea is made from the dried and fermented leaves of the *Camellia sinensis*, a camellia plant that grows in Asia. There are also no family ties to the cabbage palm *(Coryline australis)*, which is called *titrea* by the Maori, the indigenous people of New Zealand.

There are other essential oils, however, that are rightly called tea tree oils. In other countries, they replace the original Australian oil from the *Melaleuca alternifolia*. They either belong to the same species (cajeput or niaouli) or the same family (Manuka). Even though ravensara is a member of the laurel family (Lauraceae), its oil has so many components in common with melaleuca oil that its oil, too, can be included under the heading of tea tree oils. The aroma specialist and research scientist Inge Andres has investigated the tea tree oil family (which, for reasons of simplification, always refers to the essential oil of the *Melaleuca alternifollia*), and she introduces the various plants, their components, and their importance in aromatherapy on the following pages. Furthermore, recommendations are given concerning the usage and restrictions of the oil.

Cajeput Oil

Parent plant:	*Melaleuca leucadendron* or *cajeputi*
Family:	Myrtle (Myrtaceae)
Found in:	Australia, Indonesia (Moluccas, Java, Malaysia)
Interesting components:	1.8-cineole, alpha-terpineol, viridiflorol

Cajeput is an evergreen tree with narrow leaves and white blossoms. The essential oil obtained from its leaves is often an interesting alternative to the original tea tree oil. Until World War II, cajeput oil was the most important of the tea tree oils throughout the world, and was exported to Singapore, Germany, Great Britain, and Holland.

It has a bright, fresh, and clear aroma, and—because of its high 1.8-cineole content—is reminiscent of

The paperlike white bark of a cajeput tree trunk

eucalyptus, although somewhat mellower and more delicate.

For this reason, it is more suitable for children and sensitive adults. The fresh aroma stimulates breathing and the body's natural defenses, so it is beneficial for all illnesses of the respiratory system and as a preventive for colds.

On a mental or psychological level, cajeput oil stimulates clear thinking and revives us when we feel a lack of energy, but without causing hyperactivity. Folk medicine of India and Indonesia also recommends using cajeput oil for the external treatment of rheumatic pains, burns, and skin inflammations. However, because its effectiveness in treating rheumatic disorders has not yet been proven, and, in some cases, irritation of the skin has been noted, Germany's Federal Ministry of Health (nowadays, called the Federal Ministry of Medicine and Medical Products) has not recommended the use of this oil in treating rheumatic complaints.

Usage

Put a drop of cajeput oil on a ball of cotton or a handkerchief, and breathe in deeply. You can also add cajeput oil to your shower gel or body lotion; to make it more gentle on the skin, add a few drops of lavender oil or geranium oil to it.

Restrictions

Use cajeput oil carefully during pregnancy, and apply to the skin only sparingly.

Niaouli Oil

Parent plant:	*Melaleuca viridiflora*
Family:	Myrtle (Myrtaceae)
Countries:	Madagascar, Australia, New Caledonia
Interesting components:	Pinene, viridiflorol, 1.8-cineole

The evergreen niaouli tree has pointed, needlelike leaves and a flexible trunk. Around 40 percent of the French Pacific island of New Caledonia is covered with this tree.

The odor of the essential oil is clear and light, and is reminiscent of eucalyptus with a suggestion of pine. A slightly strange, tangy dimension added to this gives niaouli oil its unique aroma.

Niaouli oil is very gentle to the skin.

Of medicinal importance is its efficacy in combating such disease-causing agents as fungi and bacteria. In contrast to cajeput oil, niaouli oil is very gentle to the skin, so it is better suited for bathing or soaking the skin or parts of the body. Furthermore, niaouli oil is used to ease coughs, clear up the respiratory system, and stimulate the immune system. And viridiflorol, one of its components, is said to be similar to estrogen in its effect on the body.

Usage

Inhalation, hydrotherapy, irrigation, baths, local skin treatment.

Restrictions
Not suitable during pregnancy or for children.

Manuka Oil

Parent plant:	*Leptospermum scoparium*
Family:	Myrtle (Myrtaceae)
Location:	New Zealand
Interesting components:	Beta-caryophyllene, cardine, geraniol, linalool, geranylacetate, pinene, leptospermone

Manuka is a robust evergreen shrub that can quickly grow as tall as a tree. Its leaves are narrow and hard, and it has luxurious blossoms in the spring. Manuka is the traditional remedy of the Maoris of New Zealand, where it is harvested on the East Cape. Its essential oil has a pleasant scent, and it is an effective alternative to the original tea tree oil of Australia.

The aroma of the oil is very complex: apart from a flowery rose fragrance, it also has a woody component with a spicy base. However, compared to melaleuca oil, its odor is softer and warmer, yet the medicinal properties of manuka oil

Luxurious pink blossoms of the manuka shrub

are very similar to those of melaleuca: manuka oil is antibacterial, fungicidal, and antiviral, and it is even more effective in combating yeast fungi than the original tea tree oil.

In addition, manuka oil relieves many skin ailments, such as athlete's foot and eczema. It's also important to mention the influence of this essential oil on the psyche: manuka strengthens the nerves, improves mental balance, has a calming effect, and increases the body's natural resistance to disease.

Usage

The ways it is used are as numerous as the components of the oil's aroma. Manuka oil is suitable for inhalations, for general use on the skin, and as an additive to skin-care products.

Restrictions

Manuka should not be used in a pure form, but always mixed with carrier oils.

Ravensara Oil

Parent plant:	*Ravensara aromatica*
Family:	Laurel (Lauraceae)
Countries:	Madagascar, Réunion
Interesting components:	1.8-cineole, pinene, terpineol acetate

The ravensara tree, which can grow to a height 60 feet, can also be considered a member of the tea tree family. The name comes from the indigenous people's words *ravina* (leaves) and *tsara* (good) and thus means "good leaves." This laurel tree has a certain similarity to the European laurel tree, but the leaves of the ravensara are larger. French aromatherapists look upon ravensara as being just as multifunctional and indispensable as lavender.

The aroma of ravensara oil is light, fresh, and clear, with a woody note that has a soft, sweet background to it. It helps us to breathe deeply and freely. In contrast to eucalyptus and laurel, ravensara has a warm undertone, which rounds off its freshness and makes it more agreeable. French aromatherapists frequently make use of this oil in the treatment of shingles and influenza. Furthermore, it helps as an expectorant in the expulsion of mucus and it activates the immune system.

In terms of its mental and psychological effects, it strengthens the nerves and improves concentration. Ravensara oil activates people who are fatigued mentally, and is also used to combat insomnia. This may seem paradoxical, but clearing the mind helps us to fall asleep more easily, and sometimes we are actually too tired to fall asleep, so a light stimulant is then needed. Ravensara also balances out frequent mood changes with its nerve-strengthening properties. And in combination with lavender, it works wonders in alleviating stress.

Ravensara oil strengthens the nerves and improves concentration.

Usage

In cold poultices, and as an additive to baths or in skin-care products.

Restrictions

None are known.

Tea Tree Oil Research

Laboratory Tests

However much we may wish it were so, tea tree oil is not a miracle cure, even if we sometimes have the impression that it's a panacea, or a cure-all, because of its many areas of use. A great deal of these uses have been examined and their effects confirmed to a certain extent by researchers; other areas of use, however, are still in a speculative stage, yet seem reasonable because of various components that are also present in other essential oils. But it's important to understand that tea tree oil is in no way a magic potion, because its effectiveness can easily be explained by taking a look at its components.

Panaceas were at their height of usage during the Middle Ages and were produced in many different varieties by the alchemists of the day. Today, such products enjoy sales in the millions, but they are more likely to be called names like "wonder weapons." The word "panacea" was coined from Panakeia, the name of the Greek goddess known as the great healer. Panakeia's father was Asklepios, the god of medicine, and her sister was Hygieia, the goddess of health.

By and large, the possibility of tea tree oil as a single remedy curing so many complaints is due to five of its properties, none of which, however, has been investigated sufficiently:

- antiseptic (effective against bacteria, fungi, and viruses),
- local anesthetic,
- cooling,
- stimulates circulation, and
- stimulates the growth of new cells.

In the following, we will see to what extent these effects have been proven.

No Chance for Bacteria

The Australian Aborigines can be credited with the earliest use of tea tree oil to kill bacteria or to stop their growth. However, in the 1920s, Arthur R. Penfold was the first to systematically test and document the antibacterial property of the oil. In many laboratory tests, Penfold analyzed and compared the effect of tea tree oil and other essential oils with phenol, the most important disinfectant at that time. Because phenol is poisonous to the human organism, its use is no longer permitted. Even then, there were doubts about using this disinfectant, which severely irritated the skin, and therefore scientists were looking for an alternative.

In the 1920s, the antibacterial effect of tea tree oil was tested in the laboratory for the first time.

Research scientists were pleased with the test results of tea tree oil for two reasons: first, tea tree oil proved to be 11 times more effective against salmonellae than phenol, and, second, it didn't cause any irritation of the skin after use. But, moreover, it was

Tea tree oil was found to be an ideal combination: powerfully antibacterial yet gentle to the skin.

bound by the protein found in blood and pus and thus inactivated.

During the past decade, scientists throughout the world have tested the many possibilities for using this essential oil. In so doing, they discovered that tea tree oil kills a great number of those bacteria that cause illnesses (a selection is shown in the table below). In addition, they found that only a low concentration of the oil is needed, and that the antibacterial properties of the oil depend on the oil's composition.

It's important to point out that bacteria do not always cause illnesses. In fact, many different kinds of bacteria live on healthy skin and by so doing form a protective acidic barrier, thereby making it more difficult for microorganisms causing disease to penetrate the body. But what if tea tree oil is also detrimental to the normal flora of the skin? Considering that an increasing number of skincare products contain tea tree oil, this question would seem to be a legitimate one.

In 1996, Australian scientists gave the all-clear sign regarding this concern when they discovered that pathogens react far more negatively to tea tree oil than do the bacteria of the normal skin flora.

As gratifying as these research findings may be, it doesn't necessarily follow that the effect on the human body will be the same as the results of investigations made in the laboratory. They can only be taken as first points of reference and will have to be confirmed by clinical tests (which will be discussed in greater detail later).

Some of the Pathogens Tea Tree Oil Resisted in Laboratory Tests

Pathogen	Infections (Examples)
Staphylococcus aureus	Boils, abscesses, suppurating infections of the oral cavity
Streptococcus faecalis	Infections of the urinary tract
Streptococcus pyogenes	Cuts and scrapes, severe inflammation of the mouth and throat
Propionibacterim acnes	Acne
Escherichia coli	Infections of the bladder and kidneys
Pseudomonas aeruginosa	Disorders of the respiratory tract, pyelitis, ear infections, infections from cuts and burns

Antifungal Effects

The efficacy of tea tree oil in combating fungus has not been documented as extensively as some would want us to believe. The only fungus that has been thoroughly tested so far for the fungicidal effect of tea tree oil is *Candida albicans*. This fungus causes many infections of the skin and the mucous membranes, especially in the mouth and the vagina. Because this yeastlike fungus can be grown easily on culture mediums, the effect tea tree oil has on it can almost be seen with the naked eye, and "Australian Gold" proved to be very potent in combating candida. Other fungi, like those of the mold family (Aspergillus), and the tinea fungus (Trichophyton), which causes athlete's foot, have not been tested as extensively. But in these cases, too, tests carried out in the laboratory have shown that tea tree oil is an effective antifungal agent.

In a theoretical comparison of tea tree oil with other antiseptics, drawn up and published in extracts by the Arts and Crafts and Science Museum in Sidney, the Australian essential oil came out best because of its wide range of effectiveness. Reference antiseptics were various forms of alcohol, phenols, aldehydes, and chlorhexidine, as well as compounds of chlorine, iodine, mercury, and quatenary ammonium. Tea tree oil not only fights a great number of bacteria but also spores and fungi. None of the other disinfectants could show such a broad range of use. Unfortunately, in actual practice, very high concentrations of the essential oil are necessary in order to disinfect as thoroughly as conventional antiseptics (necessary, for example, to disinfect the instruments used in a medical operating room).

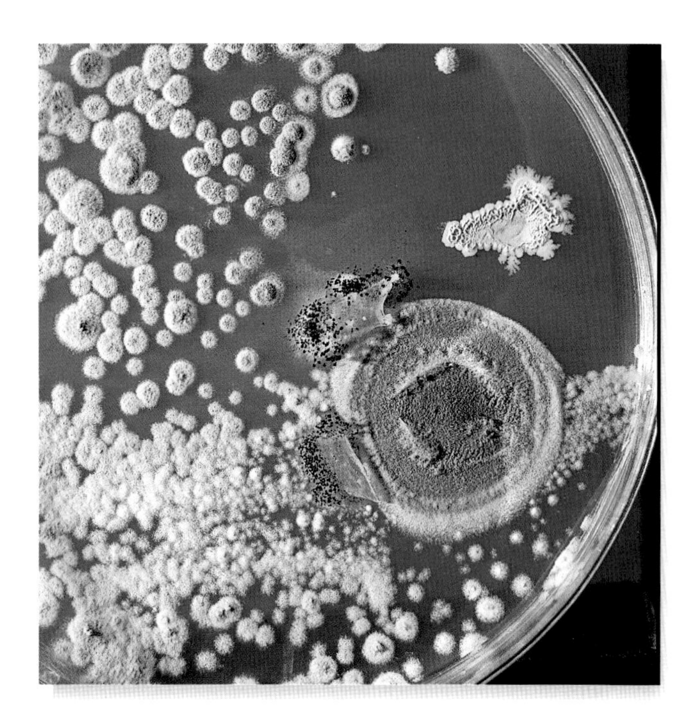

In laboratory tests, tea tree oil has proved effective against many pathogenic bacteria.

Can a Virus Be Stopped, Too?

As soon as physicians believed they had all diseases under control after the discovery of antibiotics, a new, submicroscopic infective agent, the virus, started to cause problems. Unlike bacteria, viruses don't have a metabolism of their own but lead a parasitic existence in their host, such as the cells of plants, animals, or people. A treatment with antibiotics, which specifically destroys the vital components of bacteria and thereby stops them from multiplying, doesn't work with viruses.

Viruses cannot be bred in cultures the way bacteria can, because these parasites can only grow in living cells, making laboratory tests more difficult. Another issue is the great adaptability of the virus, as seen in the annual outbreaks of influenza, for example.

It is not certain yet whether tea tree oil can fight viruses.

These properties are the reason why the successes achieved in treating bacteria have not been able to be matched when dealing with a virus. Tea tree oil will certainly not be able to do much in combating AIDS, influenza, or jaundice. But it is possible that such viral diseases of the skin as herpes simplex (cold sores), shingles, or warts can be healed with the oil. At present, however, there is more speculation than knowledge concerning the influence of this essential oil on viruses.

Cautious optimism does seem to be in order, however. Chris Bishop and his colleagues at John Moores University of Liverpool have given us the first leads in showing that tea tree oil can stop a virus. In the *Journal of Essential Oil Research*, they reported on their tests with the tobacco mosaic virus, which is especially harmful to tobacco. Plants that were sprayed with a diluted solution of tea tree oil were less affected by the virus than untreated plants. Furthermore, it was found that this protection could be increased by a higher concentration of the oil. Although these encouraging test results cannot simply be applied to the viral complaints of human beings, they do provide motivation for further research.

Additional Useful Properties

Tea tree oil has quite a few other attributes apart from those tested in the laboratory. Rather than being revealed from clinical studies, they were discovered and often confirmed from practical use. One of the many possibilities of use, as reported from experience, is local application in cases of pain, such as from cuts, insect bites, or toothaches. But no one has taken the trouble so far to do any research into the reasons for this application. It is known, however, that some of the other essential oils do have an anesthetic effect, which can be partly explained by the cooling caused by the evaporation of the essential oil.

Tea tree oil not only cools, but it also improves the circulation and thereby has a warming effect, as well. At first, this may seem like a contradiction, but it isn't. The constituents of this essential oil take effect deeply inside the skin due to the ability of the oil to penetrate the skin.

The Australian surgeon Humphrey was not the only physician to find that his patients' wounds healed quicker when treated with tea tree oil. However, in an experiment using rabbits, the treatment of wounds with tea tree oil showed neither a faster nor a slower healing.

Clinical Studies

If laboratory tests of a substance are successful, it doesn't necessarily follow that the corresponding medicine will prove to be a success in actual practice because there are considerable differences in many areas. Therefore, the effect on humans will have to be examined in clinical studies. Dosage, decomposition, and dispersal throughout the body, as well as tolerance to the medication, will have to be taken into account. However, none of the clinical tests of tea tree oil has as yet been faced with such stringent criteria as the random selection of the test persons. The necessary money for this type of testing is difficult to raise due to the relatively small tea tree oil industry. Many of the tests have been nothing more than observations made by physicians during treatment, whereas other tests have at least been carried out using a placebo (an innocuous substance) or a standard medication in what is called a double-blind test (in which neither the doctor nor the patient knows which medicine has been used).

The Ideal Antiseptic?

In 1930, the surgeon Humphrey published one of the first surveys of clinical studies in the *Medical Journal of Australia*. It was a summary of his experience with 35 percent tea tree oil for use in therapeutic treatment.

Humphrey used the essential oil for cleaning and disinfecting wounds, and afterward covered them with a dressings soaked in tea tree oil. The result was impressive: the pus disappeared, and the surface of the wound was cleaned completely without any damage to the adjoining tissue. He also used tea tree oil successfully for infections of the fingernails and toenails as well as for sore throats. In addition, he praised its exceptional disinfecting properties for the doctor's office.

Almost at the same time, a number of dentists discovered this plant remedy. Tea tree oil was used successfully as a disinfectant before and after treatment of the teeth and the gums. Furthermore, dentists recommended the essential oil as a remedy for bad breath (halitosis), aphthea (small cankerous boils in the mouth), and daily hygiene of the mouth and the teeth. In 1930, MacDonald wrote in an Australian dental journal that tea tree oil came close to being the ideal antiseptic for dentists.

Tea tree's essential oil is an antiseptic with many uses.

Seven years later, Penfold and Morrison presented a wide range of ailments that they had healed with tea tree oil. This included frostbite, cankerous boils, gynecological disorders, skin infections, skin rashes caused by bacteria, ringworm, athlete's foot, disorders of the throat and pharynx, and infections of the gums.

Healing Boils

Encouraged by positive laboratory tests, Henry Feinblatt, a doctor from New York, tested tea tree oil as a remedy for boils. In most cases, these were deep, cankerous inflammations that were caused by staphylococcus. The essential oil was used on 25 out of a total of 35 test persons.

Already after the first week, the symptoms had either nearly or completely disappeared in 24 patients in the tea tree oil group (96 percent success). The lancing of the boil was necessary with one patient only. In comparison, there was no improvement in the boils of the 10 untreated persons. In fact, five patients had to have their boils lanced. In summarizing the results, Dr. Feinblatt recommended always trying the traditional treatment with tea tree oil first before operating.

Clinical studies have shown the effectiveness of tea tree oil in treating vaginal infections.

An Aid for Vaginal Infections

Dr. Eduardo F. Pena was the first to study the effect of tea tree oil in gynecological complaints. This gynecologist concentrated mainly on trichomonal vaginitis and *Candida albicans* infections. In his gynecological clinic in Miami, Dr. Pena discovered that most of the vaginal infections of his patients were caused by trichomonads. He associated these infections with the use of unclean sanitary napkins and tampons. Trichomonads are not bacteria but one-celled flagellated protozoans that are parasitic in many animals. There were 130 women suffering from various types of vaginal infections who participated in his clinical study.

In more than three quarters of the cases, trichomonads were the cause of the complaint. In some cases, the pathogen had advanced as far as the cervix. Only four of the women were infected with the fungus *Candida albicans*. Dr. Pena wanted to find the proper concentration that would be sufficient to kill trichomonads as well as the candida fungus. For this reason, he used different solutions of tea tree oil with his patients in both vaginal douches and tampons. At the end of a six-week treatment, it was shown that a 0.4 percent tea tree oil solution was just as effective against trichomonads as the traditional vaginal suppositories were.

In 1985, Professor Paul Belaiche, the director of the Phytotherapy Department at the University of Paris, carried out a series of studies to find out if vaginal infections caused by *Candida albicans* could be cured with tea tree oil. Earlier, this French medicinal plant research scientist had used other essential oils, such cinnamon oil, oregano oil, mint oil, and thyme oil, in the fight against microorganisms. Apart from thyme oil, however, they all caused severe irritation of the mucous membranes, which meant that they were not suitable for treating vaginal disorders.

Because *Melaleulca alternifolia* was so well tolerated and had shown to be effective against candida in laboratory tests, Professor Belaiche was encouraged to perform clinical studies with 28 women who were suffering from *Candida albicans* infections. For three months, the patients inserted a small suppository containing 20 milligrams of tea tree oil into their vaginas before going to bed. Apart from one patient who stopped the treatment because she experienced a burning sensation in the vagina, none of the other women had any side effects. Twenty-three patients, 82 percent of the group, were completely healed of such symptoms as itching and vaginal discharge. In three quarters of the cases, fungus cultures were no longer found either. This was reason enough for Professor Belaiche to come to the conclusion that tea tree oil is one of the most important essential oils, as well as one of the best weapons in homeopathic aromatherapy treatment, in fighting bacteria and fungi.

Relief for Foot Problems

In 1972, an American doctor, Morton Walker, studied the effect of tea tree oil on various foot problems with 60 patients. He not only chose three different forms of the oil—a 40 percent tea tree oil emulsion, an 8 percent tea tree oil ointment, and the pure oil—but he also varied the methods of application: sometimes the problem areas were painted with the preparation or massaged with it once or twice a day; in other cases, the feet were bathed once a week. At the end of the therapy, the positive outcome was that 58 of the 60 patients were cured or at least had improved substantially.

It is still not clear whether tea tree oil can cure athlete's foot or just eliminate its symptoms.

Dr. Walker emphasized in particular the outstanding results that were achieved with sweaty feet, which are not just an annoyance with regard to odor but also a basis for getting athletes' foot. The preventive effect with athlete's foot was quite obvious even with patients who were chronically troubled with this infection. The itching, redness, burning sensation, and peeling of the skin disappeared with those patients already infected with athlete's foot. In rare cases only were pathogens still found. Even with corns, calluses, and inflammation of the ball of the foot, there was improvement.

However, the most recent studies on athlete's foot (1992) have surely been somewhat of a disappointment for tea tree oil fans: Australian research scientists compared the effect of 10 percent tea tree oil creme with the antimycotic medicine tolnaftat and a placebo. After applying them for a month, all 104 participants were tested for fungus. Although the fungus had cleared with 30 percent of the patients in the tea tree oil group, this percent was just a little higher than that in the placebo group! Tolfonat, however, had an 85 percent rate of success. Scientists assumed that the 10 percent dosage of tea tree oil was too low or the application in the form of a creme was not suitable in this case. To save the reputation of tea tree oil, it must be said that such symptoms as itching, skin rash, inflammation, and burning sensations disappeared in 65 percent of the patients who had used this essential oil.

Even Dr. Walker reported at the beginning of the 1970s that pure tea tree oil was not very effective in treating fungus of the toenails. The results of David S. Buck and his team were similar. In 1994, they tested "Australian Gold" in some parts of the United States as a remedy for the Trichophyton fungus. For comparison, about half of the more than 100 patients who had been suffering for more than a year from this fungal infection of the nails were given the substance clotrimazol, a standard medication against fungal infections. Although the six months of therapy with the essential oil cured more patients than clotrimazol did, the actual healing success was only 18 percent. But at least 60 percent of the symptoms of both groups had disappeared for good.

Therapeutic Value with Acne

An Australian research team assumed that due to its disinfecting properties, tea tree oil had to be effective in treating acne. At the same time, the question arose as to how the homeopathic medicine would be tolerated by the skin. In 1990, in order to clarify the benefits and risks of its use with acne, research scientists compared the essential oil for the first time with benzoyl peroxide, a substance that had already proven itself in combating this disfiguring skin condition. At present, benzoyl peroxide lotion is the most widely used medication in treating the less serious forms of acne.

During the three-month period of treatment using a 5 percent tea tree oil preparation, the doctors examined 124 test persons regularly and documented changes in their skin with regard to oiliness, redness, scaling, itching, and dryness. Furthermore, the patients were asked whether they had noticed any side effects and, if they had, which ones.

To judge by the results of the tests, medium to severe forms of noninflammatory acne can be treated just as well with tea tree oil as with benzoyl peroxide. However, with the inflammatory forms, the results of the essential oil were not quite as good as those of the comparison substance. In both cases, however, a bit more patience is required with treatment using tea tree oil. As far as side effects are concerned, tea tree oil was far superior to the standard medication. Although 44 percent of the tea tree oil group complained about slight side effects, there were nearly double as many in the comparison group. Scaling of the skin, dryness, and itching—typical side effects at the beginning of a therapy using benzoyl peroxide—did not occur at all with the patients using tea tree oil.

The positive benefit-to-risk relationship of tea tree oil is another reason to use it as an alternative to standard preparations in the treatment of slight to moderately severe cases of acne.

Careful cleansing of the skin is fundamental to every acne treatment.

How Does Tea Tree Oil Work?

A Comparison with Antibiotics

Tea tree oil has no highly specific effect in the way it works, which is why it cannot be called an antibiotic but only a homeopathic antiseptic or disinfectant. Antibiotics, however, make a direct attack on the structure or the metabolism of the bacteria—in other words, they "pretend" to be a component of the bacteria's cell wall. The bacteria die when they fall for this trick and accept the wrong component. Various antibiotics restrict the building up of the important components of a bacteria cell. But because antibiotics do not harm the body's cells, they can be used on the skin and mucous membranes and, like most disinfectants, be taken internally.

Even antibiotics are natural in origin. Today, most of the time these compounds used in therapeutic treatments are slightly changed chemically. Because antibiotics are so highly specific, they often lead to resistance in the body.

Tea tree oil, on the other hand, damages bacteria cells in an unspecific way, which means that the attack is not specialized for the metabolism of the bacteria. As far as we know, tea tree oil attacks the cell membrane and the cell wall, finally destroying the protein inside the cells. Due to this unspecific effect, it cannot be ruled out that the essential oil can also damage the human cell if it is used in large quantities. The most important antibacterial component of tea tree oil is terpinen-4-ol, which makes up more than 35 percent of the total. The other substances have more of a synergetic effect, which means they support and strengthen the effect of terpinen-4-ol, for example, by reinforcing its ability to penetrate the bacteria cell.

However, bacteria are "able to learn," or, to put it more correctly, are very adaptable due to their extremely rapid changes from generation to generation. The genetic changes brought about by this are the reason why they are able to produce substances that can paralyze an antibiotic, or to change the structure of their cells in such a way that antibiotics can no longer be effective. Furthermore, a bacteria cell gains in resistance through the exchange of information that it can pass on to other bacteria.

The unspecific mechanism of tea tree oil has the advantage of not being able to be so easily outwitted or bypassed by a bacteria cell. Therefore, it is not to be expected that resistance to tea tree oil can build up. An increasing number of doctors have started advising against treatment with antibiotics and instead are prescribing antiseptics with more frequency, especially for the local treatment of the skin.

It is unlikely that bacteria will become resistant to tea tree oil.

Tolerance of This Essential Oil

External Application Is Safe

As with other essential oils, tea tree oil burns when it comes into contact with the eyes. That's why use for an eye infection or with a sty (an inflamed swelling of a sebaceous gland at the margin of the eyelid) is not recommended. (Don't use chamomile tea on the eyes either, because this can cause an allergy.) Otherwise, tea tree oil was described in most tests, even the earlier ones, as being well tolerated by the skin. Results that appear contradictory at first can usually be explained by the differences in the concentrations of the oil used in the experiments.

For example, clinical studies on the skin—with 28 participants, conducted over a period of three weeks using varying concentrations of tea tree oil and a placebo—showed only a very slight reddening of the skin when a 10 percent creme was used. In another study, a 5 percent gel preparation was well tolerated by 61 test persons.

It follows that tea tree oil in a concentration of less than 10 percent—when used externally—is a well-tolerated and safe antiseptic. Neat, or undiluted, tea tree oil, however, should only be used for a short time as a first-aid treatment, and never over large areas of the skin.

People who have an allergic reaction to other essential oils should not only consult their physician before treatment, but they should also obtain information about the substance from a pharmacist, aromatherapist, or aromatologist.

Does Cineole Irritate the Skin?

In spite of tea tree oil usually being well tolerated, skin reactions have been reported more and more frequently. In the meantime, the search for the possible cause has begun, and 1.8-cineole (eukalyptol) was quickly thought to be responsible for the skin irritation.

Cineole has very beneficial medicinal properties and has therefore been used successfully in treating illnesses of the respiratory tract for some time already. Only a short time ago, the substance caused a sensation at a symposium on homeopathic medicines. There, among others, the latest research results on the effect of cineole were presented: it was said to be very antispasmodic and anti-inflammatory in high dosages. A positive therapy result similar to that of cortisone, which was used for comparison, was achieved after a few days of treatment for bronchial asthma.

The Search for the Offender

A monograph on the raw materials in perfumes written by Opdyke did not mention any allergic reactions when using cineole. A team of authors, consisting of Ian Southwell, Susanne Freeman, and Diana Rubel, confirmed this in an impressive way: they said they were unable to detect any negative skin reaction even when cineole was used in a 28 percent concentration, twice as high as that permitted by the Australian Standards Association.

Knight and Hausen brought forth a further "piece of the puzzle" that helped to rehabilitate cineole. These scientists tested 1 percent solutions of various tea tree oil components on seven patients who had had allergic reactions to tea tree oil: in this case, as well, there was no reaction to 1.8-cineole! However, six patients were allergic to d-limonene, five to alpha-terpinene and aromadendrens, and one each to terpinen-4-ol, para-cymene, and alpha-phellandrene. The test showed that limonene was the most frequent allergen.

But here, too, be careful not to make any hasty decisions, because scientists at the Thursday Plantation laboratories discovered that para-cymene can also cause severe skin irritations. This substance, which is usually present in tea tree oil in small percentages only, through the influence of light and oxygen creates from terpinen-4-ol, alpha-terpinene as well as gamma-terpinene, and through improper or too lengthy storage can develop into as much as 30 percent of the essential oil.

Scientists and tea tree oil producers are presently trying to trace the offender, which indicates that an effort is being made to quickly find the substance that triggers the intolerance reactions. Already negative publicity concerning the side effects of tea tree oil are threatening to cause a collapse of the booming market. In order to protect users from damaging their health and to preserve the good reputation of "Australian Gold," scientists are at present trying to eliminate possible allergens by selective breeding, new distillation methods, and gene manipulation.

The possible allergens in tea tree oil have not been identified beyond all doubt.

Not Recommended for Internal Use

If taken internally, tea tree oil can cause irritation of the mucous membranes and allergic reactions.

There is great difference of opinion regarding whether or not tea tree oil should be taken internally. Whereas in French aromatherapy, it is quite common to take even larger amounts of essential oils internally, Australian experts disapprove of internal use because they cannot rule out irritation of the mucous membranes in the digestive tract as well as the possibility of allergic reactions.

In fact, some incidents have been reported when tea tree oil was taken internally. In most of the cases, the quantities taken were far too high and therapeutically not customary. In two of the cases, when two children took 25 milliliters of tea tree oil by mistake, this caused diarrhea and an overall feeling of sickness, although these symptoms disappeared within 24 hours. In another example, someone experienced a state of confusion after ingesting 10 milliliters of the oil, but this passed within 5 hours. One doctor reported on the unique case of a 60-year-old man who got a "dramatic skin rash" while experiencing a general feeling of sickness after taking half a spoonful of tea tree oil. Earlier treatment with the same amount of oil, however, had shown no negative side effects. It is possible that the patient had become sensitized after taking tea tree oil repeatedly and that finally an allergic reaction was the result. But it could just as well have been impurities in the essential oil that caused the side effects.

The concentrations of tea tree oil causing symptoms of poisoning in tests with animals were never higher than concentrations of other essential oils that are approved for internal use by the health authorities in Germany. They definitely recommend the internal use of a few drops of eucalyptus oil, pine-needle oil, or fir-needle oil. Furthermore, these oils are basic components of several cough syrups. In addition, in Germany, preparations can be purchased in soft gelatin capsules containing high doses of substances obtained from essential oils.

Surely tea tree oil cannot do any harm to adults if it is used with the proper precautions. The question is, of course, for what effect it should be used. Because the greatest part is breathed out from the lungs, it most certainly should have a healing effect on the respiratory tract. But even here it is recommended to use only approved medicines. The much propagated intestinal rehabilitation, or antifungal, course of treatment (if at all necessary) cannot be achieved with just a few drops of the oil, because not enough of the substance actually reaches the intestines. As long as scientists have not given the green light for internal use of tea tree oil, children, at least, should not be treated at all, and adults only under the guidance of a physician or an experienced nonmedical practitioner.

Overall Safety

A study called for by the Australian Tea Tree Industry Association (ATTIA) and the New South Wales Department of Business and Consumer Affairs in 1989–90 examined the safety of tea tree oil by testing it on animals. The individual tests, conducted according to international regulations, showed among other things that:

- tea tree oil if taken internally is not more toxic for rats than eucalyptus oil; this finding is from a test called the comparable oral toxicity of essential oils;
- no sensitization occurred with guinea pigs, even though the outer layers of their skin had been injected repeatedly with the oil;
- the use of undiluted tea tree oil on healthy as well as damaged skin of rabbits did not cause irritation; and
- the essential oil did not cause any change in genes, nor did it cause cancer (mutagenicity test, according to Ames).

The present research findings on effectiveness, quality, and harmlessness of tea tree oil are to date insufficient for the oil's approval as an internal medicine in Germany. In the meantime, however, the essential oil is being taken seriously because of the great demand for it. In 1996, it was included in the German Pharmacology Code, which is a supplement to the *German Book of Medicines*. Here, the standards are set with regard to the minimum quality of the parent (original) plants and their testing as pertains to components and falsifications. As in the *Australian Standards*, the German Code maintains that tea tree oil may be produced only from *Melaleuca alternifolia*, *Melaleuca dissitiflora*, and *Melaleuca linariifolia*, and should contain at least 30 percent terpinen-4-ol and a maximum of 15 percent 1.8-cineole. Of special importance are the regulations with regard to proper storage, which, if they are adhered to, can prevent changes in the oil and certain side effects.

The German Pharmacology Code provides standards for tea tree oil.

IMPORTANT

Perishability and Storage

As with all essential oils, tea tree oil has to be stored in a cool, dark place and in full containers, if possible. Then it remains stable for approximately a year after opening. Otherwise, warmth, light, and air can break down important substances into other products (oxides)—for example, terpinen-4-ol into para-cymene, which is suspected to be the cause of skin irritations. Also, be sure to keep tea tree oil—like all other medicines—out of the reach of children.

Self-Treatment with Tea Tree Oil

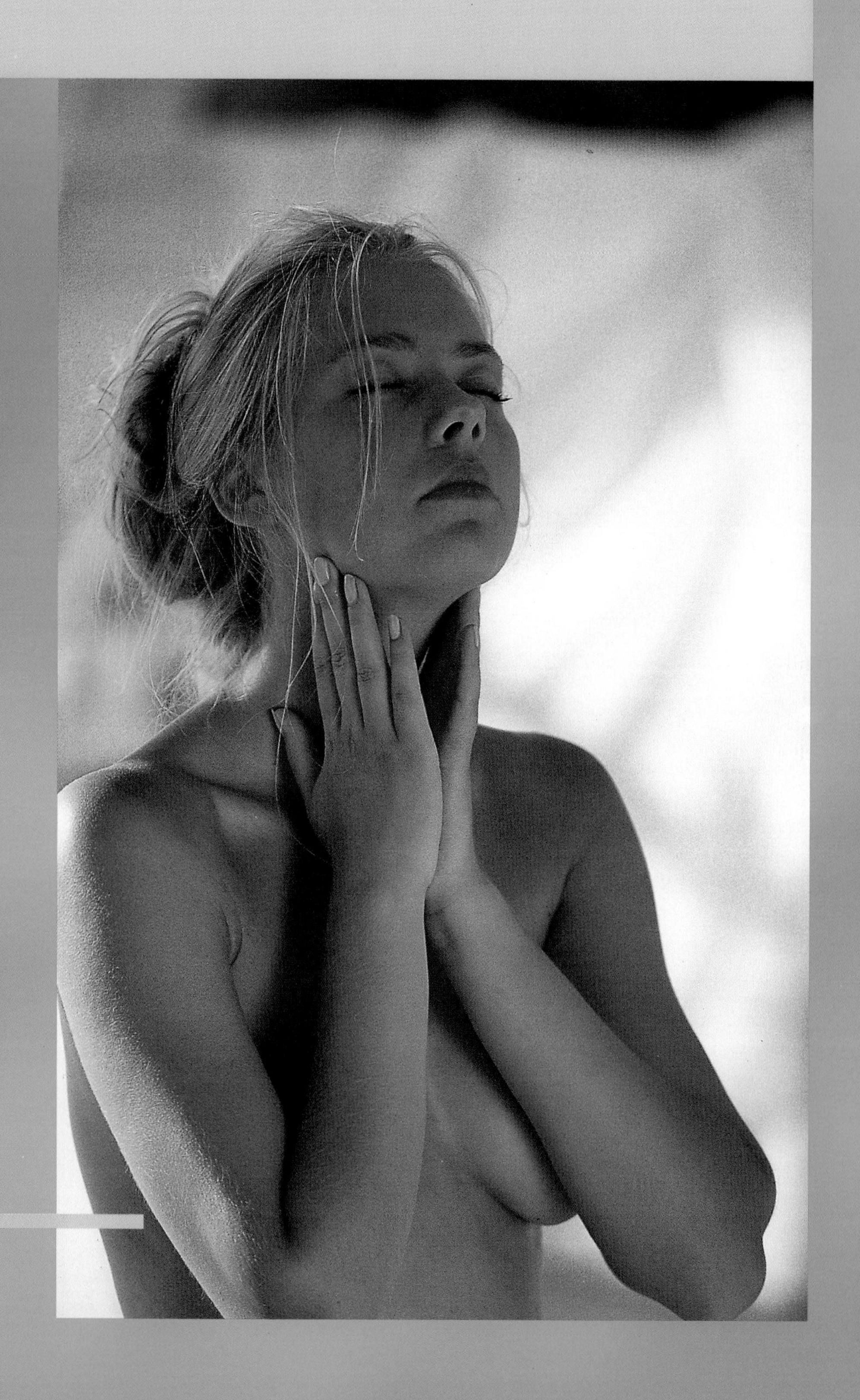

Oil, Hydrolate, and Solutions

Tea tree oil is usually applied externally, which means on the skin, hair, and nails, but it may also be applied directly to certain mucous membranes (for example, in the mouth and nose) if it is not to be taken internally. Neat tea tree oil is used only in special cases and then only at very particular points: for instance, with fungus of the nails that is very difficult to reach or to swab a single pimple, but even in such cases a dilution is often recommended. Under no circumstances, however, should larger areas of skin be treated with the concentrated oil, because more likely than not it will lead to skin irritations.

Tea tree oil does not dissolve very well in water, so it follows that it is difficult to mix the oil with water. Here are some useful tricks that can help:

- You can buy ready-mixed solutions of 15 percent tea tree oil in water that can be further diluted to any extent. In addition, you can buy a synthetic emulsifier, which is a diluting agent between the oil and water phases.
- You can also prepare a solution of tea tree oil and water yourself. A suitable emulsifier is polysorbat 60 (Tween 60®). It is, of course, easier to have your pharmacist prepare the solution properly and then dilute it further according to your specific needs.
- Or you can make an emulsion of 40 parts of tea tree oil to 60 parts of 13 percent alcohol (isopropyl alcohol, which you can get at your local drugstore), and you can then dilute it further.
- An emulsion using the natural emulsifier cholesterol from milk or cream should be used immediately. For this, you simply stir tea tree oil into whole milk or cream and add the mixture to your bath water. This preparation is suitable for foot and sitz baths and for vaginal douches.

A little milk helps to dilute the oil more easily in water.

Generally, it is recommended to prepare only small amounts of a solution, because its durability is much less than that of the pure essential oil.

You can mix tea tree oil with all kinds of fatty oils; high-quality oils,

OUR TIP

Plastic syringes and measuring spoons, which you can buy cheaply at your drugstore or pharmacy, are very suitable for measuring out ingredients exactly in the mixture.

A small wine-testing glass or a small measuring cup is another practical as well as inexpensive aid for measuring. However, in most cases, it is not necessary to measure out the amounts to the exact milliliter.

such as almond oil, wheat germ oil, avocado oil, and jojoba oil, are especially suitable. In recent years, jojoba oil—really a liquid wax—has been used frequently in cosmetics, because it is so well tolerated by the skin. This oil, which comes from an olivelike fruit found in Mexico, California, and Arizona, has other advantages as well: it does not turn rancid, it is very heat-resistant, and, despite being a natural product, it is nearly free of all impurities.

You can also buy tea tree oil hydrolate, which is the water left over as a by-product during the distillation of the essential oil. It contains traces of the essential oil and the water-soluble substances of the tea tree leaves insofar as they also entered the distillate. You can add the hydrolate to your bath water or apply it directly as a lotion for the face or hair.

OUR TIP

In general:
1 drop of oil is equivalent to 0.03 milliliters (ml)
1 ml of oil is equal to 30 drops
Example:
If you want to make a 40 percent tea tree oil solution, put 4 ml of oil (120 drops) into a small measuring cup and then add 10 ml of the previously mentioned alcohol.

Areas of Use from A to Z

Although tea tree oil can be used in many ways, it shouldn't be mistaken for a cure-all. Furthermore, only a few of the suggestions for use are supported by scientific research. The remainder are based on personal experience. Although it is fairly certain that the essential oil can help a variety of ailments, it must be said that some of the claims of healing belong more to the realm of myth. However, on the following pages, you will find instances in which tea tree oil can truly help you, as well as how you can use it.

This practical part tells you about tea tree oil preparations that you can—in combination with other substances, as well—prepare yourself or buy ready-made. Additional tips will tell you what else you can do to alleviate your complaints. When using tea tree oil, it's important to consider that so far there are no exact recommendations of dosage based on scientific research. To the contrary, the dosages given are based on experience only.

Generally, for each case of self-treatment, if the symptoms have not improved considerably after three to four days, you should consult your physician.

Abscesses

Abscesses are localized collections of pus surrounded by inflamed tissue in the deeper parts of the skin that can develop into painful knots. These acute, sometimes chronic infections are caused by bacteria, mostly stahpylococci. Although these pathogens also live on healthy skin, they can cause this inflammatory disorder if the defenses in the body of the patient are weakened; sometimes abscesses also occur in connection with small injuries or if the sebaceous glands are blocked.

Usually, the formation of pus develops around a hair follicle; barber's rash, or sycosis, and furuncles, or boils, are examples of such inflammation of the hair follicles, or folliculitis. Furuncles are deeply penetrating abscesses that appear on the outer part of the skin in the form of painful, hard knots, up to the size of a walnut. Their yellow centers often burst open after a few days so that the pus can drain away. Although they usually appear where hair is growing, they also show preference to places where clothes rub the skin, such as on the neck, the armpits, the shoulders, between the legs, and on the buttocks. If several boils merge, this is called a carbuncle.

Depending on the size, position, and accompanying symptoms (sometimes fever) of the abscess, you should consult a physician. Boils on the face in particular can cause serious complications that could even result in meningitis.

Use of the Oil

Nowadays, abscesses are usually treated with antiseptics or antibiotic ointments, and only in rare cases must the boil be lanced. Dr. Henry Feinblatt, an American physician, proved that good healing results could also be achieved with tea tree oil (see page 30).

Tea tree oil can be very effective in treating minor abscesses, such as barber's rash and inflammations caused by ingrown hairs. If the essential oil is well tolerated, the inflamed spot is dabbed three times a day either with neat tea tree oil or in a 10 percent solution. Warm compresses soaked in tea tree oil bring additional relief. To do this, you soak a piece of sterilized cotton in hot water and wring it out; then you add 3 to 4 drops of tea tree oil, and place it on the affected skin for about 10 minutes.

Put warm tea tree oil compresses on boils.

Acne

A pimple showing up on a Saturday night is the horror of every teenager. And young people often withdraw into themselves completely because they are worried about being ridiculed when they get a severe case of acne. In such cases, acne is more than a cosmetic problem and therefore a physician should be consulted.

Acne during puberty is not just a cosmetic problem.

Even today, many myths surround this skin disease, which should be taken seriously and usually appears during puberty but can last up until about the age of 30.

Acne during puberty is generally a genetic problem and has three phases. In phase one, the cornification of the skin is characteristic, and results in a constriction of the pores. During puberty, when more hormones are produced, the production of sebum, a fatty lubricant matter, increases, and it cannot drain off due to the constricted pores (phase two). This in turn produces comedones, or blackheads. Finally, during phase three, the *Propioni acnes* bacteria, which finds an ideal culture medium in the sebum, decomposes the blackhead and thus causes inflammation.

The medical therapy is carried out according to the severity of the acne. In serious cases, vitamin-A acids (which must not be used during pregnancy, because they seriously damage the fetus) and antibiotics are prescribed. Apart from this, taking birth-control pills can improve the skin of many women. Often pimples disappear as soon as the hormonal changes stop after puberty.

Other Helpful Measures

- Acne diets are superfluous, according to the experts. This doesn't mean, however, that "junk food" is good for the skin. A well-balanced and wholesome diet not only keeps you fit but also reflects itself on your skin. That's what we mean when we say: "You are what you eat." However, many people believe that it's not necessary to completely give up your french fries and chocolate just because of your pimples.
- Nonetheless, you may be allergic to certain foods, which you then have to avoid. It is quite possible to be allergic to such healthful foods as carrots, or perhaps strawberries, just to mention two examples.
- If you have problems with pimples, you should not smoke, because this not only constricts the tiny blood vessels in your face and causes poor circulation in the skin, but the "blue haze" itself contains substances that are harmful to the skin.
- Even if your acne is not primarily caused by improper hygiene, cleansing, disinfecting, and sometimes peeling of the skin are some of the most important measures you can take to prevent pimples or to treat them. After all, bacteria, cornification, and excessive sebum production are what cause this condition.
- Inferior cosmetic products can also bring about acne due to their comedogenic ingredients that can block the pores. If your skin is prone to have blackheads and pimples, make sure that your cosmetics do not contain such substances as lanolin, cocoa butter, olive oil, peanut oil, coconut oil, polyethylene glycol, sesame oil, coal tar, and stearates.

A healthful diet is an important means of internal skin care.

Tea tree oil is gentler to the skin than other substances for the treatment of acne.

Use of the Oil

As one team of research scientists (see page 32) discovered, tea tree oil is very suitable for the treatment of light to moderately severe forms of acne: it penetrates the skin easily and is effective against the bacteria causing the acne, but at the same time it is less aggressive to the skin than other acne preparations. Skin care has been greatly facilitated by the many cosmetic products you can buy with tea tree oil in them. It's recommended that you proceed as follows in your daily cleansing and care of your face:

1. Clean your face using only alkali-free soaps or cleansing lotions. Use only cotton pads or throwaway facecloths, because normal washcloths provide breeding grounds for bacteria. Always rinse off all the soap carefully with lukewarm water.
2. Afterward, apply a face toner containing tea tree oil; use cotton pads to do this.
3. Finally, rub a light moisturizing cream containing tea tree oil on your face.
4. Individual pimples can be disinfected in between the cleansing periods by dabbing them with tea tree oil. In addition, it's helpful to do a face-peeling treatment twice a week.

It's important to leave the removal of blackheads to your dermatologist, otherwise disfiguring scars could result. If you find that you just have to remove blackheads yourself, do so very carefully after a facial steam bath to which a few drops of tea tree oil have been added or after applying warm compresses by pulling the skin around the pimples apart with your index fingers (don't squeeze!). To avoid infection, first wrap your fingers in clean facial tissues. Afterward, the skin should be disinfected.

You can also easily make up tea tree oil preparations yourself—for example, by adding a few drops of tea tree oil to your cosmetic products, which, of course, have to be suitable for problem skin. What follow are a few suggestions for making your own cleansing and facial-care preparations:

- Cleansing: Mix 10 drops of tea tree oil with about 10 milliliters of your usual cleansing product, or mix 15 drops (with the aid of a little milk) with 10 milliliters of water. Remember that the emulsion should always be freshly prepared. Furthermore, don't save at the wrong end, because too low a concentration of tea tree oil does not disinfect properly.
- Toning: For this, a mixture of 15 percent ethanol (50 ml) and 30 drops (or 1 ml) of tea tree oil is suitable.
- Disinfecting individual pimples: If no skin irritations have occurred, you can apply neat tea tree oil with a cotton swab directly to the affected parts of the skin.
- Steaming the face: Fill a small bowl with about a pint of boiling water, and add 10 drops of tea tree oil. Then put a towel over your head,

close your eyes, and steam your face for 5 to 10 minutes.
- Disinfecting after the proper removal of individual pimples (see above): Dab the spots with a mixture of 70 percent ethanol (50 ml) and 30 drops (or 1 ml) of tea tree oil.

Athlete's Foot

There has been a definite increase in fungal infections recently, and mainly the feet have been affected. Nearly 20 to 30 percent of the entire population suffer from athlete's foot. A lack of hygiene, but sometimes also excessive hygiene, can be the culprit. The protective acidic covering of the skin is destroyed by soaps that dry out the skin, as well as by daily showers. Fungus can penetrate the skin through the damaged protective skin cover. Fungus can also penetrate easier if the skin has been softened. Therefore, it is easy to transfer a fungus from foot to foot in public swimming pools, saunas, and showers. That's why it's not surprising that there are a great number of infections among athletes and soldiers who frequently shower together. People whose bodily defenses have been weakened are also endangered: this is the case with people who have diabetes mellitus, serious infections, or cancer or are undergoing treatment with medicines that suppress the immune system (such as cortisone).

Here are some ways to avoid athlete's foot:

- Always wear bathing slippers in public swimming pools, saunas, and showers, as well as in hotels.
- Dry your feet carefully after washing.
- Avoid wearing polyester socks and shoes not made of leather; don't wear rubber boots or tennis shoes the entire day.
- Change your socks and shoes every day.

The correct diagnosis is important before starting treatment of athlete's foot, and the diagnosis is usually done by the afflicted people themselves. Characteristic symptoms are a reddening of the skin, scaling, and a burning sensation, or a whitish discoloration of the skin with scaling. People with a weakened immune system suffer more frequently from athlete's foot.

You should consult your physician when large areas of the skin are affected or if you discover a lot of dry scales or whitish skin. Fungal infections of the feet usually take a long time to cure, and the treatment is not always successful. Even though the symptoms may get better a few days after the feet have been treated with antifungal medication, the therapy should be continued for several weeks. Because the antimycotic preparation initially kills all the active forms of the fungus, the spores themselves are not injured. Thus, new fungus can develop from these spores if the conditions are favorable.

Even under normal circumstances, that means without any complications occurring, treatment should last for

For athlete's foot, the feet are sprayed with a highly concentrated solution of tea tree oil.

A regular foot bath with essential oil helps to prevent athlete's foot.

about four weeks. The earlier and the more thorough the therapy, the higher are the chances of a cure. The following measures should accompany treatment with antifungal preparations:

- Take care of your foot hygiene, which means washing your feet and drying them thoroughly before applying an antifungal preparation. Many dermatologists even recommend blowing the feet dry with a hair dryer.
- Change your towels and your socks daily during treatment; the bed sheets should also be changed frequently.

- Whenever possible, wear open, loose shoes without rubber soles. Change your shoes regularly.
- If you suffer from sweaty feet, it is recommended to place small strips of gauze between your toes.
- Patients with metabolic disorders, such as diabetes, can improve their blood circulation and thereby their immune system by doing foot exercises and massaging their feet.

Use of the Oil

The efficacy of tea tree oil in treating athlete's foot is still disputed, because the results of the clinical tests have been inconclusive (see page 31): although tea tree oil was able to clear up symptoms like itching, burning, and scales in nearly all cases, fungi were still detected.

As far as the necessary dosage is concerned, no specifics can be given yet. However, tests show that a 10 percent tea tree oil creme was not sufficient to bring about a cure (however, it is not clear whether it was due to the concentration of the essential oil in the creme or to the way in which the creme was prepared).

If you tolerate it well, you can try a highly concentrated solution (30 to 50 percent), which you can apply with a cotton pad, or you can spray it on (you can get small atomizers at your local drugstore). If the athlete's foot does not improve during the first week or if the symptoms return after the therapy is terminated (apply the solution for four weeks at least!), it will be necessary to use an antibiotic.

With athlete's foot as well as with sweaty feet that promote athlete's foot, warm foot baths with tea tree oil are recommended, as well. Afterward, you should rub a foot creme containing tea tree oil onto your feet and in between your toes (also see page 81).

Burns, Minor

The cooling of burns is most important, which means that large areas have to be held under cold, running water for at least 15 minutes. When doing this, you should be careful not to cool the person down too much—especially children. Afterward, it's advised to consult a doctor.

Use of the Oil

With minor burns, the cold water takes the immediate pain away. Later, a longer-lasting cooling effect can be

Pure tea tree oil is dabbed onto minor burns.

achieved by dabbing tea tree oil onto the burn, because its local anesthetic properties go to work directly on the nerve endings, which in turn brings about the desired soothing effect. If a blister opens, tea tree oil can be used to prevent an infection of the wound from bacteria.

Tea tree oil (neat) should only be applied to smaller burns. Dab it carefully and directly onto the blister with a cotton swab.

Chicken Pox and Shingles

Chicken pox and shingles have the same pathogenic agent, a virus called *Varicella zoster*. Chicken pox starts with a fever and a few vesicles (small blisters on the skin filled with liquid). The vesicles spread from the trunk over the entire body, but they usually don't affect the hands and feet. Although this illness is relatively harmless and rarely produces complications, the tormenting itching provokes scratching, which leads to an inflammation of the blisters and can cause scarring.

Easing the pain is the most important measure when treating shingles.

After a chicken pox infection, which usually occurs during childhood, the *Varicella zoster* virus lies dormant in the body for years, and only later in life when the immune system is weakened does it become active again. Then the pathogen, which, by the way, belongs to the herpes viruses, attacks the nerves of the spinal cord; on the areas through which the nerves pass, it causes a painfully blistered rash, which spreads like a belt in a semicircle around the waist. That's why the Germans call it "rosy belt." The English word for this illness, "shingles," stems from the Latin word *cingulum,* which also means belt. Younger people don't come down with shingles very often, but if they do, it is not so painful for them as for an adult.

Use of the Oil

Both illnesses are infectious and spread by an airborne virus. Although you can hardly influence the course of either one, the patient should be looked after by a doctor.

With shingles, therapy to alleviate pain is especially important, and should be supported by high dosages of vitamin B. In addition, a bath to which tea tree oil has been added can disinfect the open blisters, thereby protecting the patient from a bacterial infection and reducing the itching.

Mix about 20 drops of tea tree oil with some cream or milk as an additive for a bath. (The patient should have a good constitution for such a bath.) The pustules can also be dabbed with a mixture of the essential oil with almond oil or jojoba oil (20 drops in 20 ml).

Nurses have obtained good results when treating patients suffering from shingles with a 5 to 10 percent tea tree oil lotion, which they applied three times a day to the affected skin areas.

Colds

Inhalations with essential oils can bring relief if you have a runny nose. Ear, nose, and throat specialists recommend them, especially as an accompanying therapy for long-lasting or chronic infections of the sinuses. When the steam is breathed in, the essential oil reaches into the farthest corners of the respiratory tract. Although this doesn't affect the cold virus very much, at least it makes it easier to breathe through the nose, because the swelling of the mucous membranes there is somewhat reduced and the entire head becomes clearer. Most people find it more pleasant to use tea tree oil for inhalations instead of peppermint oil, because the latter stings the nose considerably. Incidentally, chamomile is an ideal supplement to tea tree oil.

Use of the Oil

A scent lamp, an aroma stone, or a vaporizer can be used to fill the room for some time with the vapor of tea tree oil, which will have a beneficial effect on your general well-being. In addition, you should inhale the vapors directly once a day, preferably in the evening in the following way.

Bring a quart of water to a boil, and pour about half of it into a plastic inhaler. Make sure that the inhalation mask covers your mouth and nose only, and not your eyes. The electric steam diffusers (some drugstores will lend them to you) are ideal for this, as are inhalers whose water pot is coated with a Styrofoam sleeve so that the water will be kept hot longer. Now, add 5 to 6 drops of tea tree oil to the boiling water (don't add any milk or emulsifiers) as well as 10 drops of chamomile extract, and inhale for about 5 minutes. Then bring the rest of the water to a boil again, add that to the inhaler, and inhale for another 5 minutes.

Today, plastic inhalers are available that cover only your mouth and your nose.

Colds—Prevention

The primary areas where essential oils have always been used are with colds, because it is here that their disinfectant properties work best.

Bacteria and viruses are the main causes of coughs, runny noses, and influenza. Sometimes we "catch a cold" very easily, whereas at other times everyone around us can have a cold and we remain healthy. Thus, our immune system is responsible for our becoming ill or not.

Our immune system is weakened when we:

- eat an unbalanced diet and lack vitamins,
- are constantly under pressure or stress,
- are always discontented and unhappy, or
- ingest an excess of such poisons as alcohol, cigarettes, and certain medications.

Before viruses and bacteria enter the bloodstream and come into contact with our body's defenses, they have to get past the mucous membranes that function like a natural barrier. Entry is easier when the mucous membranes are very dry—for example, when we stay too long in heated or air-conditioned rooms—or if our blood circulation is impaired from our being too cold.

Use of the Oil

Not only vitamins but also natural substances from plants, such as *Rudbeckia hirta*, a perennial composite herb with showy flowers, are helpful in improving the immune system. Tea tree oil is also thought to have a positive effect on the body's defenses, although to date no clinical tests have been conducted.

Tea tree oil should not be taken internally as a preventive but should be inhaled instead. Add 3 to 4 drops of the oil to the water in an aroma lamp or a diffuser or vaporizer. You can experiment with this process, because no sides effects are expected from inhaling.

The best ways to prevent a cold are to strengthen the immune system, keep the mucous membranes moist, and ensure good blood circulation throughout the entire body.

Coughs

If you have a cough, you should also see to it that the membranes of the nose and the throat are kept moist. The measures mentioned under "Sore Throat and Hoarseness" (page 71) apply here, too. However, it is even more important to make up for the liquids lost. That's why you should drink more than usual when you have a cold; warm tea and juices are best. Furthermore, inhalations are recommended, because they clear the respiratory tract.

We know that a component of tea tree oil, 1.8 cineole, which is also found in eucalyptus oil, increases the movement of the cilia in the bronchus. This eases the clearing of mucus from the respiratory tract. It is not certain whether the essential oil liquefies the mucus as well, but in any case it eases coughing. Again, it is important to drink enough liquids.

IMPORTANT

Please note that essential oils must not be used on infants' faces, because this can lead to a sudden swelling of the membranes of the respiratory tract or breathing standstill reflex.

Use of the Oil

Apply the following mixture to your back and chest in the morning and especially before going to bed.

For adults:

Tea tree oil	10 drops
Pine-needle oil	10 drops
Almond oil	20 ml

For children:

Tea tree oil	3 drops
Pine-needle oil	3 drops
Anise oil	3 drops
Almond oil	20 ml

All the ingredients have to be mixed well and applied lukewarm.

Cuts and Abrasions

Cuts and abrasions occur every day. We frequently get cuts when working in the kitchen or the garden. Abrasions on the knees and the elbows, so common in children, are perfect places for harmful bacteria to enter the body and are often a cause of infections. So that nothing more than perhaps a small scar remains, apart from disinfecting the wound it is imperative that a tetanus shot be given, something we are often inclined to forget, especially when it comes to adults. However, a tetanus shot doesn't give you life-long protection; it only lasts for about 10 years, and then you have to have a booster shot. It is necessary to get a tetanus shot immediately if the wound has been in contact with possible carriers of pathogens, such as the soil, wood, rusty nails, grass, and hay, as well as being bitten by an animal.

Use of the Oil

In spite of the protection provided by vaccination, the wound has to be thoroughly disinfected, which can be done perfectly well with tea tree oil in the less serious cases. Laboratory tests show that the antiseptic effect of this essential oil is not reduced by blood or any secretions from the wound. Furthermore—unlike many other antiseptics—irritations of the skin occur only rarely after treatment with tea tree oil. Another special advantage is the slightly anesthetic effect of tea tree oil on the wound, which at the same time eases the patient's pain.

Small cuts and abrasions can be cleaned with a 10 percent tea tree oil solution (the solvent being 70 percent isopropyl alcohol) and/or swabbed with neat tea tree oil. However, larger injuries should be treated by a doctor. And remember protection against tetanus!

In addition, a report made by a nurse at the Munich City Hospital, which appeared in an aromatherapy

Small cuts or abrasions can hardly be avoided when working in the garden.

journal, is worth mentioning. After a patient had surgery for a fractured ankle, she treated the wound—which hadn't healed for weeks and therefore amputation was being considered—with various essential oils. For 15 minutes every day, the patient took a foot bath to which was added 3 drops of tea tree oil, citrus oil, and thyme oil (white), as well as a teaspoonful of salt. Afterward, the wound was dabbed dry with sterile compresses, covered first by an oleic gauze with 3 drops of tea tree oil on it, and then dressed with a sterile compress and a gauze bandage. After three weeks, the wound had healed completely!

Dandruff

The entire skin of the body is subject to a constant renewal process: when new skin cells move up so that the upper layer of the skin has to give way, this layer then dries up, dies, and finally flakes off. This is a normal process that also takes place on the scalp. Generally, we don't notice this when it is happening, because the small scales disappear down the drain when we wash our hair or they stick to our towels and bed linen. Only if the keratinization of the skin cells takes place too quickly do the scales float down onto our clothes, where they become increasingly noticeable and irritating. Often this is accompanied by a severe itching of the scalp. If a lot of sebum is produced at the same time, the scales form clumps and stick to the scalp, and if you have dark hair this can be an extra nuisance.

Men—especially between the ages of 15 and 25—are more often affected by dandruff than women are. Sometimes dandruff quite simply gets worse because of certain hair-care products, including shampoo, hair spray, permanent-wave solution, and hair dye, which can irritate the scalp. This means that when you stop using these products, your dandruff and the itching will disappear, too. Dandruff can also be caused by neurodermatitis or psoriasis, which can crop up on other areas of the skin, like the elbows and the knees.

In most cases, dandruff is caused by an excessive production of sebum, for which genetic or environmental factors are thought to be responsible. If, in addition, the protective acidic covering of the scalp is damaged, a good breeding ground for a certain type of fungus called *Pityrosporum ovale* can develop, and it in turn stimulates cell growth and the flow of sebum.

Use of the Oil

Apart from actively fighting the yeast fungus by means of the substance ketoconazol, antiseptics also bring about relief, which is why we can assume that tea tree oil can be used with success, too. Unfortunately, positive results from research are still lacking, but many people with dandruff have reported that their dandruff has been reduced through treatment with tea tree oil.

The yeast fungus *Pityrosporum ovale* can be the cause of dandruff.

At this point, we don't know whether the essential oil only eliminates such symptoms as itching and reddening of the skin, or if it also kills the fungus.

Use only a mild, alkali-free hair shampoo that does not irritate the scalp. Put 4 drops of tea tree oil onto a blob of the shampoo. Ready-made, mild tea tree oil shampoos are also suitable for daily hair care. A few drops of the neat oil should be added, however, for severe cases of dandruff.

What Else You Can Do

Experience has shown that dandruff improves after sunbathing. Scientists assume that the ultraviolet radiation inhibits the growth of the fungi.

Fungal Infections of the Vagina

The misconception that fungal infections of the vagina are a venereal disease is fairly widespread and makes this affliction even more of a taboo subject. Even when women are among themselves, they rarely talk about this problem although three out of four of them have suffered from it. In most cases, the fungus returns again and again. Contrary to what some people think, you do not get a vaginal fungus by a lack of hygiene or from "committing adultery." It is a commonplace, usually harmless infection that you can generally treat yourself.

Women need to be able to recognize the symptoms of fungal infections of the vagina and to distinguish them from other, more harmful infections. The symptoms of a fungal infection of the vagina are rather obvious: in addition to an increased yellowish-white discharge, which usually is crumbly and has a slight yeasty smell, the outer areas of the vagina are affected by itching, reddening, and in most cases also swelling.

Usually women recognize these symptoms when they appear repeatedly, and they often treat them with chemically based antifungal preparations, which can be bought at a drugstore without a prescription. If the symptoms don't improve after three days, a gynecologist should be consulted. You should also go to your doctor under the following conditions:

The flora of the vagina are built up again with the aid of lactic acid bacteria.

- if this is the first time you have suffered from a fungal infection,
- if you are less than 18 years old,
- during pregnancy, because there could be other causes for the complaint,
- if infections appear more than four times a year, or
- if the infection is accompanied by a fever, bad odor or a grayish-green discharge, overall feeling of sickness, blisters in the genital area, aching of the lower abdomen, or difficulties when urinating; these symptoms may indicate an infection of bacteria or viruses or trichomonads.

As already mentioned, women don't just pick up this fungus; it is nearly always present even in a healthy vagina. The normal flora of the vagina are mainly composed of bacteria that produce lactic acid, the lactobacilli, which ensure that there is a slightly acid milieu in the vagina that fungi do not like and that prevents them from spreading. However, if this sensitive ecosystem becomes out of balance, the fungi can gain the upper hand.

Women suffering from recurrent bouts of fungal infections of the vagina should pay special attention to the following points:

- These infections are encouraged by excessive hygiene: perfumed soaps, additives to the bath water, vaginal douches, and vaginal sprays damage the flora in the vagina when used to excess.
- Panty liners, as well as panty hose, made of nonporous materials can cause a concentration of heat and moisture (the "greenhouse effect"). The same holds true for underwear made of synthetic materials and for jeans that are too tight. It is better to wear cotton or silk panties, change them daily, and wash them in hot water.

- Chemical gels, foam, or suppositories used for contraception can also harm the bacterial flora.
- Hormone fluctuations also affect the vaginal milieu. The pH-value of the vagina is increased and its moisture content is reduced by birth-control pills (the high-dosage generation) and during pregnancy, menstruation, or menopause, and both encourage the growth of fungi. Due to a lack of vaginal moisture, the mucous membranes no longer have a protective film, and there are less lactic bacilli, whose production of acid would otherwise suppress the fungus. Women with little vaginal moisture should not wear tampons during the last days of menstruation, because they dry the vagina out even more and thereby make it more susceptible for fungal infections.
- Antibiotics not only destroy disease-causing bacteria but also useful microorganisms. This is why when you take antibiotics, the flora of the intestines and the vagina are affected detrimentally.
- A weakened immune system, such as when you have a cold, an inflammation, or a lack of vitamins, is an ideal condition for the growth of fungi.
- Medicines that suppress the immune system, like certain drugs for treating cancer or cortisone, clearly decrease the body's defenses against fungi.
- Patients suffering from certain metabolic diseases, especially diabetes, are more susceptible to fungal infections.
- Presently, there is a great deal of speculation about the effect food has on fungal infections. At any rate, a wholesome diet with a moderate-to-low sugar content is recommended.

Fungal infections of the vagina are not a venereal disease.

Use of the Oil

Treating a vaginal infection naturally with tea tree oil calls for a little more patience than when using the chemically based antifungal preparations. But in many cases, it is just as effective, and can also be tried if the chemical substances are not tolerated very well. Furthermore, tea tree oil often functions as a knight in shining armor when traveling or on weekends. It is, however, always advisable to go to a gynecologist who can check the success of the treatment by means of a smear test.

Two clinical tests have not only confirmed that tea tree oil is effective in treating fungal infections of the vagina, but also that it is tolerated very well by the healthy flora of the vagina (see page 30). German gynecologists, who are usually receptive to natural healing methods, often treat vaginal infections alternatively with essential oil.

Lyall Williams, a researcher of tea tree oil who compared the effectiveness of various essential oils on the yeast fungus *Candida albicans* in a laboratory survey made in 1995, judged the value of tea tree oil as follows. Although synthetic antifungal preparations fight candida more effec-

A treatment with antibiotics can encourage a fungal infection of the vagina.

OUR TIP

You can work out the diluting factor (x) with a simple formula:

$$\frac{\text{Concentration of the basic mixture (in percent)}}{\text{Concentration of the final mixture (in percent)}} = x$$

Example: a 20 percent tea tree oil solution to be made from an 80 percent one.

$$\frac{80}{20} = 4$$

This means that the basic solution has to be diluted with up to four times the amount of water (alcohol).

tively than tea tree oil does, he maintained, it needs to be considered that many vaginal infections are not caused by fungi alone. In such cases, he continued, tea tree oil makes a favorable impression by its broad range of effectiveness against fungi, bacteria, and trichomonads, and its good tolerance.

In the United States, women are now able to purchase tea tree oil suppositories to treat vaginal fungi. In many places elsewhere, women have to use slightly more cumbersome methods. (The dosages recommended here are based on medical experience obtained by trial and error in various doctors' offices.)

For the basic tea tree oil mixture, combine tea tree oil and 13 percent alcohol (from your druggist) in a proportion of 4:6. This emulsion is diluted to 1:40 for hip baths, ablutions (use throwaway washcloths only), and vaginal douches. Ready-made solutions, which you can dilute with boiled water down to a 1 percent solution, are just as effective. For hip baths, the basic diluted mixture (about a pint) is, of course, further diluted by adding warm bath water.

A tampon is soaked in a 4 percent tea tree oil solution (the basic mixture diluted 1:10), and is changed twice a day.

Some women even use concentrated tea tree oil and put a few drops (4 to 6) on a tampon. However, this can cause irritations of the mucous membranes in the vagina. According to the experience of some women, a dilution of tea tree oil with fatty oils, such as almond oil, in a proportion of 1:10 or 1:20 is more suitable. The red oil of Saint-John's-wort, which is rich in tannic acid, is suitable as a carrier oil for a mixture in which to soak tampons: simply dilute 1 ml of tea tree oil with 20 ml of Saint-John's-wort oil. The treatment of your partner, or at least very careful hygiene, is also recommended, even if your partner doesn't have any symptoms of a fungal infection. This avoids what is known as the Ping-Pong effect, namely reinfection. Although any natural treatment of a fungal infection of the vagina requires a lot of patience, the itching and burning as well as the discharge should have clearly improved after a week at the latest; if not, a gynecologist should be consulted.

Following natural treatment as well as after using a "chemical cannon," it is important for the flora of the vagina to be built up again as quickly as possible. Lactic acid bacteria are suited best for this purpose, and you can get them in capsule or tablet form at your local drugstore. Experts advise against using tampons soaked in yogurt, because not only can the desired lactic acid bacteria enter the vagina but unwelcome germs can, too.

An internal yogurt cure, however, cannot do any harm and is always worth trying (unless you have an intolerance to milk). A comparative survey made in the United States showed that women who ate more than a cup of yogurt every day were protected better from vaginal infections than women who did not eat any yogurt.

OUR TIP

Instead of diluting tea tree oil with water, it is recommended to use an infusion made with oak-bark tea for hip baths. Its tannic acid relieves the itching and helps to soothe inflammation. Pour some cold water onto 3 spoonfuls of oak bark (from your pharmacy or health food store), and heat it to boiling. Leave it to steep for 5 minutes, strain it, and add it, together with approximately 10 ml of the basic tea tree oil mixture, to a hip bath, and then fill the tub up with warm water.

Fungus of the Nails

Fungi that hide under the nails of the fingers or the toes are especially stubborn. As with athlete's foot, public places where there is a great deal of water or dampness are possible sources of infection. But in addition, there has to be increased susceptibility or very favorable conditions for the fungus in order for the nails to become infected. The following are favorable conditions for fungus of the nails:

- There is an injury to the nails or the cuticle, such as when the cuticle has been cut back.
- Rubber boots, sneakers, or rubber gloves are worn for several hours a day.
- Your shoes are too tight.
- You wear artificial nails for a long time. The nail plate rises slightly due to leverage, thereby making it easier for fungus and bacteria to get in. Furthermore, if your manicurist doesn't work hygienically and carefully, a fungal infection can develop between the real nail and the artificial one.
- You suffer from diabetes, so the circulation of the blood in the toes is restricted.
- Your immune system is weakened or suppressed by such substances as cortisone or other medical preparations used in treating cancer or AIDS.
- You are more than 60 years of age.

Fungi can slip under the front of

Tea tree oil is impressive because of its broad spectrum of effectiveness and tolerance.

Tea tree oil is only used as a supplement to other therapy in treating fungus of the nails.

the nail or enter over the nail bed at the sides or at the back, and they can even breed on an injured nail surface. In the beginning, you can recognize a fungal infection by the yellowish coloring of the nail. As it proceeds, the nail plate vaults up, swells, and becomes brittle, until finally it peels off the nail bed. The worst case can be a complete destruction of the entire nail.

If you think you have fungus of the nail, you should have your nails checked immediately by your doctor. On the one hand, your doctor has to make sure that the changes in the nail are not caused by something else; certain chronic diseases of the lungs, heart, and liver produce similar symptoms that could be mistaken for a fungal infection by the layperson. On the other hand, the rapid and effective treatment of the infection is important to ensure lasting success.

A treatment with antifungal creme is usually not very effective, because it doesn't adhere to the nail very well and is wiped off quickly. More promising are the new antifungal preparations, which are applied like a colorless nail polish and whose substances penetrate to the affected tissue slowly and constantly. In addition, the varnish seals the nails, preventing the fungus from further spreading. Application is quite simple, but nonetheless the treatment is a rather lengthy one, because the nail usually has to grow out completely and this takes about three months.

If larger areas of the skin or even several nails are affected, your doctor will probably prescribe an additional, systematic therapy, which means you will have to take an antifungal preparation orally, but this is normally tolerated quite well.

Supplementary Use of the Oil

You should not treat fungal infections of the nails with tea tree oil alone, even though clinical studies have shown that in some cases the essential oil was successful in treating fungal diseases of the nails (see page 32) and could compete with the usual antifungal cremes. The failure quota is too high, and a delay of several weeks could decrease the chances of healing the nail or the skin. Nevertheless, tea tree oil can be used as a supplement to the nail-varnish therapy.

If you suffer from fungus of the nails, proceed as follows. If possible, cut the nail plate down to where it is attached to the nail bed. Next, file away any sharp edges and the nail plate itself with a throwaway nail file. Now, soak the nails for about 10 minutes in a disinfecting bath, to which you can add approximately 30 drops of tea tree oil. Then dry the nails carefully, or even better, use a hair dryer to blow them dry. Finally, apply an antimycotic nail polish thinly, according to the instructions of the manufacturer. Avoid as much as possible doing any cleaning with liquids and unnecessarily washing your hands during treatment with the fungal nail polish.

Gangrene

Patients who must keep to their beds for long periods of time usually get bedsores. Those areas of the skin under constant pressure, like the skin on the heels and the shoulders, can easily break open. When this happens, the condition is called gangrene. Because the skin of elderly people in particular does not heal easily, gangrene must be avoided by intensive care. This means that the patient's position in the bed has to be shifted frequently so that the same areas do not always have to carry the weight of the body. In addition to this, their body weight should be supported and shifted with the aid of special pressure cushions.

Gentle massaging of the sensitive areas in order to improve blood circulation strengthens the tissue and thereby prevents bedsores. If in spite of massage, sore spots develop, they have to be disinfected carefully, because germs can enter more easily at such points. In addition, ointments should be applied that encourage cell regeneration and thereby support the healing process. Dexpanthenol, the alcohol of pantothenic acid, does this, so you might look for a lotion with this ingredient.

Use of the Oil

In some hospitals and the nursing wards of some homes for the elderly, the effectiveness of tea tree oil in preventing gangrene has been successfully tested. Based on the preparations used there, the following mixture is recommended for the rubbing in of the oil:

20 drops of tea tree oil
40 ml of jojoba oil
10 ml of wheat germ oil

or

20 drops of tea tree oil in 50 ml of a lotion containing dexpanthenol

However, after running tests on the skin of rabbits, it hasn't been shown so far that tea tree oil is able to close wounds the way panthenol can. Presumably, its positive effect comes from its disinfecting property combined with a high tolerance to the oil.

Gangrene can be avoided by patients receiving intensive care whereby their position in bed is shifted frequently.

Inflammation of the Gums

Sometimes there occur acute and, in most cases, very painful inflammations of the gums or the mucous membranes of the mouth. These are caused by bacteria that are also found on the healthy mucous membranes of the mouth, but that can enter the tissue and damage it due to a weakness in the body's immune system, or a small cut or a sore spot, made by dentures, for example. Or you may merely take a "wrong bite" into the mucous membranes of the cheeks, or the tongue may get infected and heal badly.

Use of the Oil

Inflammations of the gums and the mucous membranes of the mouth can be dabbed with a cotton swab soaked in neat tea tree oil, or they can be treated with mouth rinses. Put 3 drops of the essential oil into half a glass of warm water, and rinse your mouth for about 2 to 3 minutes.

Inflammation of the Nail Wall

The nail wall surrounds the nail and protects it from the penetration of germs. Inflammation of the nail wall, which can be the beginning of a long-lasting fungal infection of the nails, is often due to incorrect and excessive nail care. Therefore, more women than men are affected, and it is nearly always a problem of the fingernails and not the toenails.

The cuticle should never under any circumstances be cut off during a manicure. At most, it can be carefully pushed back with a rubber cuticle stick or a cotton swab after having been soaked during a bath or shower. Only those bits of cuticle that have been torn may be trimmed with scissors. In addition to mistakes made during a manicure, contact with aggressive chemicals or cleaners can give rise to inflammation of the nail wall.

Tea tree oil doesn't work with every kind of skin inflammation.

Use of the Oil

Soak the affected fingers or toes for 5 to 10 minutes in a nail bath of 30 drops of tea tree oil with 10 ml of almond oil or olive oil.

Tea tree oil can also be massaged lightly into the inflamed areas.

Inflammation of the Skin

Inflammation of the skin can be caused by bacteria, allergens, or skin-irritating substances. If the pathogens responsible are bacteria or fungi, the chances of a successful cure with tea tree oil are very good. However, if the disorders are of an allergic or toxic nature, it first has to be made certain whether or not the tea tree oil itself was the cause. If this is the case, the skin reaction usually disappears by itself when the essential oil is no longer applied.

Inflammations caused by bacteria or fungi are usually due to weakened immune defenses—for example, if the natural protective acidic covering of the skin has been damaged. Because fungi spread the most in places where it is moist and warm, they prefer creases in the skin of overweight people, babies' bottoms enclosed in diapers, or the foreskin and glands of the penis.

Use of the Oil

Generally, doctors will treat inflammation of the skin with a combination of pills and ointments. However, in cases that are not too severe, it is surely worthwhile to try a tea tree oil mixture first.

In cases of diaper rash, for example, wipe the baby's bottom with a preparation of 3 to 5 percent tea tree oil in a baby oil or a fatty vegetable oil (almond or jojoba oil) base, and then apply zinc oxide ointment. Furthermore, the diaper has to be changed frequently.

Note: Never give tea tree oil to the baby internally! Inflamed areas of the skin of adults are best treated with a 5 to 10 percent lotion. This should be applied to the affected areas two to three times a day. In addition, tea tree oil can be added to the bath water or washing water.

Insect Bites

Usually we are unaware of a mosquito attack, and that's why it is almost impossible to prevent. The mosquito punctures the skin with its proboscis and injects a secretion that dilates the vessels, thereby increasing the flow of blood. In addition, the clotting of the blood is hindered, ensuring that the proboscis of the insect does not become blocked.

After the mosquito's meal is over, a white spot remains on the skin, which develops into an itchy swelling within a few hours. This is due to the injected secretion causing an allergic reaction in the human body: the body's defensive substance, histamine, is released, resulting in the reddening, swelling, and heating of the affected part of the skin. The immune system of some people reacts so strongly to the protein in the secretion that whole parts of the body can swell severely or there may be a circulatory collapse that can even result in death. Normally, however, the reaction is confined to a small area of skin around the puncture that starts to itch severely within a short period of time. If you scratch it, then you only make it worse, because the damaged cells will release even more histamine. This is a vicious circle, which you can easily break, however.

Use of the Oil

Here, a cooling gel with an antiallergic effect is suitable. It should contain an antihistamine (this stops the reaction of histamine). Before application, you can add a drop of tea tree oil to the gel.

The scientist Lyall Williams reported on relieving the itching caused by insect bites by using lotions containing tea tree oil. These lotions have been used successfully in veterinary medicine since 1942. Williams attributed this relief to a local anesthetic property of tea tree oil on the fine, exposed nerve ends.

Of course, the pure essential oil will also relieve itching if you dab it on the affected area immediately after the bite and then let it take effect.

Insect Repellent

Even if insect bites are a nuisance, most people consider them to be harmless; however, you should know that every 30 seconds someone dies as a result of them.

The danger of insect bites is often underestimated.

Hardly anyone thinks about the dangers involved if they are plagued by insects in far-away countries or even at home. The examples of malaria and Lyme disease, which is transmitted by ticks and for which there is no protection by vaccination, show how important a defense against insects and ticks is. This is how you can avoid insect bites and protect yourself against ticks:

- Always wear closed shoes, long socks, long pants or jeans, and a shirt with long sleeves, as well something to cover your head, when walking through the woods or near marshlands after dark. It's also best to wear light-colored, loose clothing.
- Always sleep under a mosquito net in the tropics, even if you have taken a malaria prophylaxis beforehand.
- Ticks should be removed immediately—if possible, with tick-pincers. Oil should not be applied in order to suffocate them.
- Protect all uncovered skin with an insect repellent.

You can enjoy your walk through the woods even more if you protect yourself against insects.

It is still imperative that you take the appropriate steps for protection by vaccination—for example, against the early-summer meningoencephalitis, which is also transmitted by ticks, as well as a malaria prophylaxis!

Insect repellents are not only an important protective measure to use in the tropics but also at campsites or at lakes or even on your own terrace at home, where insect bites—even if they aren't a danger to your life (if you aren't allergic to them)—can cause the torment of constant itching.

Use of Essential Oils

In this connection, essential oils have celebrated a comeback, partially because of the negative publicity chemically based insect repellents have received recently that link them with possibly damaging the liver and the nervous system. However, essential oils have a shorter duration of effectiveness, which means that they must be applied more frequently. There should be no side effects, as long as you don't have a contact allergy caused by any of the essential oils used.

Insect-Repellent Spray:

Tea tree oil	1 ml
Eucalyptus oil	2 ml
Clove oil	1 ml
Anise oil	1 ml

Fill up to 50 ml in a 70 percent isopropyl alcohol mixture.

This mixture can be filled in a bottle having a pump atomizer (you can buy these at your drugstore or a specialty shop for aroma oils). You can also use the small, handbag-sized perfume atomizers, but only if they are new or carefully cleaned, because otherwise the leftover perfume would attract mosquitoes.

Apply a ready-made or self-made insect repellent with a tea tree oil basis to all uncovered skin; the effect will last for about four to five hours. The insect-repellent spray above can also be dispersed in a room with the aid of an aroma fan (an electric fan for the dispersion of essential oils), which you can buy in specialty shops for aromatherapy products.

Intestinal Fungi

Fungal disorders of the intestines are much talked about these days. However, Professor Hof of Mannheim, Germany, wrote the following in a medical journal: "Often the hysteria that occurs after the fungus has been found in the intestines is unfounded." Fungi in the stool are quite normal. Disorders only occur if the number is clearly above average. Sometimes faulty testing or a delay in getting the specimen from the doctor's office to the laboratory can be the cause of finding an incorrect (that is, too high) number of fungi.

It is not only normal for fungi and bacteria to be found in the intestines, but it is vitally important because they have many protective functions. Nearly everyone who has had to take antibiotics has experienced what happens when the intestinal flora are destroyed. We suffer from diarrhea, because not only are the bacteria causing the illness killed but the useful bacteria in the intestines are killed, as well. In such a case, the fungi can spread more rapidly in the intestines, on the mucous membranes, and on the skin, because they are no longer suppressed by the bacteria. Apart from patients who have been treated with antibiotics or chemotherapy, it is mainly those persons having weak defenses against infection who suffer from intestinal fungi, such as patients who are HIV-positive, have transplanted organs, or have tumors.

Fungi usually do not have a chance to spread extensively in the intestines if the immune system is in order, because they are killed off by the white blood corpuscles in the mucous membranes of the intestines. However, if antifungal therapy is necessary, it must be carried out by an experienced physician, who will then probably prescribe a medicine for fungal infections, called an antimycotic.

The experts are at loggerheads over the value and composition of an antifungal diet.

The question of adhering to an antifungal diet is much disputed by the experts: some say that no diet at all is necessary, whereas others advise us to eat food that has a very low sugar, or carbohydrate, content. The nutrition experts of the German Nutrition Association recommend a wholesome, well-rounded diet, which

It is doubtful that tea tree oil can combat intestinal fungus.

generally will stabilize the immune system. They believe that then a moderate amount of sugar may be permitted.

As already mentioned on page 39, the internal use of tea tree oil has to date not been approved in Germany. Furthermore, a few drops of tea tree oil will surely not be effective against intestinal fungi (especially if taken on a lump of sugar, as occasionally recommended). In fact, it is doubtful whether the essential oil will reach the intestines at all, because it may be breathed out or destroyed by the stomach acids beforehand.

In addition, if concentrated essential oils are taken internally, an irritation of the mucous membranes cannot be ruled out, so we strongly advise you *not* to treat yourself internally with tea tree oil.

Lip Blisters

Lip blisters, shingles, and chicken pox have one thing in common: they are all caused by the herpes virus. The herpes simplex virus, in particular, which is responsible for lip blisters, is very widespread: it lies dormant in nearly everyone, even though it remains unnoticed in many cases. After a first infection, this virus settles down in the nerves, where it hides so that our immune system cannot find it.

Under certain circumstances—stress, for example, or intensive sunbathing or during menstruation—it shows up by causing small, painful blisters usually on the lips. Before that, however, it makes itself felt by a tightening up, tingling, and itching of the affected areas. If it is accompanied by fever or you experience pain in the facial nerves, you should see your doctor.

Nowadays, lip blisters are usually treated with antiviral substances. They are applied as soon as the lips start to tighten up or itch, and they can ease or sometimes even prevent the attack. More modern preparations that have to be prescribed by a doctor help even after the blisters have already appeared. However, all of these antiviral substances have one major disadvantage: the viruses become resistant to them very quickly so that the intervals between the attacks become shorter and shorter.

Melissa, or balm mint, a medicinal plant, is also used in the fight against herpes. But the water-solvent plant acids in it are what are thought to be responsible for its antiviral effect as opposed to the essential oil alone. Due to the special way it takes effect, resistance is not to be expected.

Use of the Oil

In practice, tea tree oil has already proven itself to be a remedy for herpes of the lips: after using the oil, patients have observed that the blisters did not spread out as much and they disappeared more quickly. So far, however, we don't know why this occurs, or whether or not the essential oil truly fights the virus.

Perhaps tea tree oil even prevents the virus from attaching itself to the

Oil baths containing gamma linolenic acid are a healthful way to care for skin affected by neurodermatitis.

cells, as does the balm mint extract. Or does it just dry up the blisters? What we do know is that tea tree oil supports treatment by easing itching and preventing an additional bacterial infection of the weakened skin cells.

If you have lip blisters, apply a little undiluted tea tree oil on the affected areas with a cotton pad or swab several times a day.

Neurodermatitis

The number of people suffering from neurodermatitis has been increasing continuously in the industrial countries of the world. So far not much is known about this illness, the main symptoms of which range from severe itching and reddening of the skin to papules and pustules on it. But experts are in agreement that its causes are in some way connected with the immune system, which could be overreacting the way it does with allergies. It's been observed that people suffering from neurodermatitis are frequently affected by illnesses that are also caused by an overreaction of the immune system, like allergic asthma and hay fever, for example.

In nearly 90 percent of all cases, neurodermatitis appears for the first time before the patient is six years of age. The risk of getting this illness is increased by the following factors:

- It is inherited, whereby the illness in the mother is of greater importance.
- The mother smoking during and before pregnancy.

- The family having higher socioeconomic standards; some causes can be excessive hygiene, stopping breast-feeding sooner, and more stress from holding down two jobs.

Neurodermatitis occurs intermittently, so there are intervals with no affliction. It's important to try to extend these intervals as long as possible, because so far no cures are known. This means that in practice everything that irritates the skin must be carefully identified, a laborious task, and then avoided.

Attacks of neurodermatitis can be caused by:

- synthetic materials that irritate the skin, but also by wool or rough and tight clothing; cotton that is washable at 100°C (212°F) is generally tolerated best;
- rapid changes of temperature, sweating heavily, or hot weather;
- showering or bathing too frequently and with water that is too hot;
- all contact allergens that come into contact with the skin; sometimes these even include substances in certain ointments used in the treatment of neurodermatitis, such as cortisone, bufexamac, and dexpanthenol, as well as various plant extracts; so, the patient, if in doubt, has to consider these substances, too, as something that might trigger a phase of the illness;
- natural allergens, such as pollen, animal hair, and house dust mites;
- food allergens; usually there are just one or two ingredients in food that do not agree with the person suffering from neurodermatits and that have to be avoided, but because these ingredients vary from person to person, there is no such thing as a "neurodermatitis diet"; and
- stress, problems at school, or family conflicts; although neurodermatitis may not be an ailment of the psyche, as the name suggests, a skin reaction can have a psychological basis.

In treating neurodermatitis, the patient should decide which remedy is best.

What is very important is taking good care of the skin, which is too dry with 98 percent of all people suffering from neurodermatitis. Due to a disorder in its upper layer, the skin is unable to retain water. That's why the skin has to be moisturized regularly and the evaporation of water has to be restricted.

The replacement of the skin oils has to be tailored individually to each patient. Here, the patient's well-being is of foremost importance. The patient shouldn't use ointments, for example, if he or she prefers not to do so, but should consider using other preparations. On the one hand, ointments are good for the skin, because they contain more oil and they prevent evaporation. But on the other hand, some patients find them so unpleasant that in acute cases they may actually promote the scratching of the skin.

In addition, oil baths at body temperature (approximately 36°C, or 98.6°F) are often beneficial, but the baths should never be too hot. These

OUR TIP

A basic rule of thumb for care of skin with a neurodermatitis affliction is moist on moist and dry on dry. This means that if acute phases are accompanied by discharges, you should apply moist compresses and cooling ointments and lotions. Contrary to this, during healthy intervals, you should use, as needed individually, the oily ointments and cremes.

baths are not meant for cleansing, but are solely to replace oil on the skin. Spreading oil baths have bath additives in them that do not mix with water and therefore coat the patient's skin with an even film of oil when leaving the tub.

Today, basic substances used in skin-care products for patients affected by neurodermatitis are the moisturizer urea, and polidocanol, which eases itching by working directly on the nerve endings that cause the urge to scratch. Use of these products reduces acute phases to such an extent that cortisone is now very rarely prescribed or, when necessary, only in small quantities.

Oils from the seeds of the borage herb and the solanum, or nightshade, plant are rich in gamma linolenic acid, and have produced good results. In some cases, the skin improved only after a few weeks of treatment, especially when the seeds were taken internally but also when they were just used in skin-care products.

Use of Tea Tree Oil

Using tea tree oil in treating neurodermatitis has some undisputed advantages:

- It works as a local anesthetic and cools the skin. That's why tea tree oil gets rid of the itching sensation, and the suffering patient no longer has the urge to scratch.
- It is also thought that bacteria are responsible for a deterioration of the skin of patients afflicted with neurodermatitis. Tea tree oil can combat the skin infections that are frequently caused by the damaged immune system.

But there is also one serious disadvantage:

- Tea tree oil can trigger an allergy. Lately, an increasing number of reports have stated that tea tree oil could cause a contact allergy.

Some doctors have had good results with tea tree oil and are in favor of its use with neurodermatitis; others, however, are very critical of it. Therefore, treatment with tea tree oil has to be conducted with extreme care. Talk about it with your doctor or a nonmedical specialist. If you tolerate essential oils, it can be used as follows:

- A 5 percent solution in cremes or body lotions can be applied. Never use neat tea tree oil!
- Add a few drops (5 to 10) to an oil bath.

- Selected skin areas can, during nonafflicted periods, be anointed with a mixture of tea tree oil with borage and solanum oils, high in gamma linolenic acid (1 drop to 20 ml).You should buy only small quantities of these oils, because they lose their effectiveness very quickly when they come in contact with the air. The mixture can also be made using almond oil or jojoba oil.
- If the skin is inflamed and red but still intact, cool compresses can be applied, because they relieve the itching and thus prevent scratching. To make a compress, soak a gauze pad or cloth in cold, strong black tea or oak-bark tea (add 2 spoonfuls of the tea to 3 cups of cold water, bring to a boil, let it steep for 10 minutes, and allow it to cool) to which you have added 2 drops of tea tree oil. Wring out the cloth slightly, and place it on the reddened skin.

Pains in the Muscles and Joints

Muscle and joint pains often occur after partaking in sports that are too strenuous or as a result of a rheumatic illness. Stiffness following sports can be avoided by doing warm-up exercises beforehand and stretching exercises afterward. Otherwise, a massage and a warm bath will ease muscle and joint pains, because the circulation of the blood will be improved, thereby easing tension.

Use of Essential Oils

This pain-relieving effect can be supported by such essential oils as rosemary, pine needle, and tea tree. Components of the Australian remedy that irritate the skin—including the monoterpenes, such as alpha-pinene, beta-pinene, carene, and limonene—dilate the smaller blood vessels. The same substances are also found in fir-needle and pine-needle oils as well as in the natural turpentine oil of various conifers. Tea tree oil also relieves pain due to its local anesthetic effect.

For your bath water:

Tea tree oil	1 ml
Rosemary oil	1 ml
Pine-needle oil	1 ml

Fill a container with jojoba oil to 50 ml, and shake well.

Massage after sports is even more effective with essential oils.

This essential oil mixture eases tense muscles and painful joints. Some companies offer ready-made massage oils with tea tree oil already in them.

Rosacea

Rosacea is a skin disease of the face that usually appears after the age of 40. Small, red veins and inflammation of the skin develop on the nose, cheeks, or forehead. These are sometimes accompanied by papules, or small knots, with fine scales, on up to festering pustules.

The cause of this disorder has not yet been fully identified. Even if an increased flow of sebum accompanies rosacea, it is not a disorder of the sebaceous glands as is acne. To the contrary, it has been found that the small knots are connected to the blood vessels in the skin.

It has been observed that rosacea appears together with certain other complaints, such as gallbladder disorders, chronic infections, genetically caused vascular weakness, and high blood pressure. Furthermore, drinking coffee, excessive consumption of alcohol or tobacco, and exposure to heat can intensify this skin disease.

Use of the Oil

First, your doctor should make a diagnosis. The use of tea tree oil (which in some cases has already proven successful) would be possible, because treatment is usually external with antiseptic or antibiotic substances. The cleansing and disinfecting of the skin with tea tree oil is similar to that in treating acne.

Cleansing: Add 10 to 15 drops of tea tree oil to approximately 10 ml of your cleanser, or mix 15 to 20 drops (with the aid of a little milk) into 10 ml of water. Remember to prepare a fresh mixture each time.

Skin toning: Make an emulsion using 15 percent ethanol (50 ml) and 30 drops of tea tree oil.

Rosacea got its name from the characteristic reddening of the face.

Sore Throat and Hoarseness

The throat should be kept as moist as possible when you suffer from a sore throat or hoarseness. During the day, you can suck sugar-free lozenges made with the disinfecting substances of tea tree oil or sage. Furthermore, it's recommended to hang towels soaked in tea tree oil in the bedroom at night. In addition, make certain that the air in your home is neither too dry nor too warm, and avoid using air-conditioning.

Disinfect your throat by gargling in the morning and in the evening (if possible in-between times as well): simply put 2 to 3 drops of tea tree oil in half a glass of warm water, take a sip, and gargle for about a minute.

Only the dentist can remove the cause of your toothache.

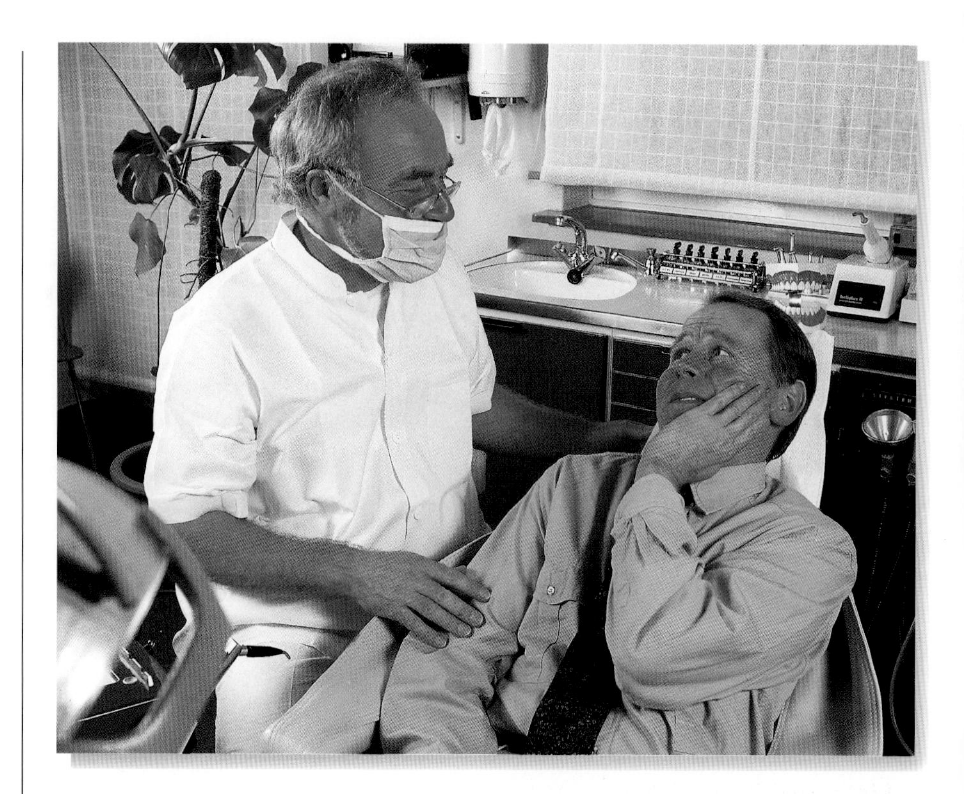

Toothache

If you suffer from a toothache, there is nothing you can do and you have to see your dentist. And you should do so as quickly as possible, because even if you take painkillers, problems of the teeth generally get worse if treatment is delayed.

Use of Essential Oils

Two proven household remedies can ease your pain until you get to your dentist. Swab the spot with tea tree oil or clove oil. Both oils penetrate deeply into the gums and anesthetize the fine nerve endings, thus bringing pain relief.

OUR TIP

Incidentally, clove oil is distilled from the spice clove, which is a member of the myrtle family and thus related to tea tree. This medicinal plant has been used in folk medicine for a long time now: if you suffer from a toothache, it is recommended to chew either the flower buds or to dab clove oil onto the gums.

Warts

Warts are benign growths on the epidermis. In most cases, they are caused by viruses, and there is a danger that the warts may spread over the body or infect other people. Sources of infection, again, are the wet floors at public pools, gyms, and so forth. Warts don't grow on the skin alone, but also on the mucous membranes, and in 80 percent of all cases they disappear just as quickly as they developed. For scientists, this apparent magic is a sign that the body' s defenses have won against the virus in the end. Warts commonly appear on hands and fingers or on the face, where they are especially obvious. They are usually quite painful when they occur on the soles of the feet.

Use of the Oil

Claims have been made that a lot of medicines and "magic potions" can do away with warts, but none of these methods has been shown to be a hundred percent successful. In some cases in treating smaller warts, tea tree oil has been effective, so this essential oil is worth a try. It has to be dabbed onto the wart frequently in an undiluted form.

Other Possible Uses

First-Aid Travel Kit

Who truly knows what to expect on a trip? Before you leave, you put together a small first-aid kit: headache tablets, adhesive tape, insect repellent—that should do. After all, the diving goggles and the bathing suit have to fit in, too. But what happens? Of course, you don't get a headache or insect bites. Instead, you injure yourself while snorkeling on a reef and get a slight abrasion, and, after sunbathing, small blisters develop on your lips. Naturally, the ointment for herpes of the lips is on your bedside table at home, and a disinfectant for the abrasion is in the medicine cabinet in your bathroom.

Due to the numerous ways it can be used, tea tree oil is valuable to take along when traveling.

Essential oils influence our feelings and sensations.

Tea Tree Oil's First-Aid Uses

- Aching muscles
- Athlete's foot
- Burns
- Colds
- Disinfecting bites and cuts
- Disinfecting the toilet
- Disorders of the feet (for example, blisters and corns)
- Fungal infection of the vagina
- Inflammation of the gums
- Inflammation of the mucous membranes of the mouth
- Inflammation of the skin
- Inflammation of the throat
- Insect repellent
- Itching
- Lip blisters
- Sunburn
- Toothache
- Vermin (lice, for cxample)

When you are traveling—especially if there is not much room in your suitcase or backpack—a small bottle of tea tree oil is an ideal companion. That's why it is called the "smallest first-aid kit in the world." Even for people for whom tea tree oil is only a second choice at home because chemical medicines seem to work much faster, the essential oil can be just what they need when traveling and they are not equipped to take care of all the aches and pains that come their way. So, whether tea tree oil is your first choice or not, it can help to relieve your complaints until you can get to a doctor.

The detachable checklist toward the end of the book lists the different ways of using the oil.

Aromatherapy

Nearly every form of therapy with essential oils can be described as aromatherapy—at least according to the French cosmetic chemist Gattefossé, who coined the word in the 1920s. He observed the antiseptic effect of essential oils and their ability to penetrate the skin and the mucous membranes, and he used them accordingly. During the esoteric movement of the 1980s, the effect of the fragrances on the senses and the psyche was in the foreground. Today, aromatherapy is looked upon as a holistic homeopathic method that can also be combined with other homeopathic procedures.

Gerhard Buchbauer, a professor at the Institute for Pharmaceutical Chemistry in Vienna, selected the following definition as a basis for this section: "Aromatherapy is the use of aromas for healing purposes, or for the easing or prevention of illnesses, infections, and indisposition, simply by the inhalation of these substances."

During inhalation, the essential oils not only reach the entire respiratory tract, but they also can be shown to be in the blood in a sufficient quantity to have an effect on the body. Moreover, the aromas stimulate the sensory cells, or the receptors, that direct the olfactory information to the brain, and so evoke sensations and feelings. A special branch of aromatherapy known as psycho-aromatherapy deals with the interaction of aromas with the psyche.

Use of the Oil

Due to the disinfecting effect of tea tree oil, inhaling it is especially suitable for the treatment of, or as an additional therapy for, acute or chronic illnesses of the respiratory tract, such as bronchitis and inflammation of the sinuses, as well as simple colds. Here, the value of this Australian remedy cannot be rated too highly. You will find examples of how to use it in the "Colds" and "Colds—Prevention" sections (page 51).

In *The Pocket Encyclopedia of Essential Oils*, the aromatherapy specialist Inge Andres discusses her experience with tea tree oil in psycho-aromatherapy. She says that it takes most people some time to get used to the odor. Its strongly herbal, tangy, and medicinal note is not very pleasant to smell, and sensitive people find it rather offensive. She explains that tea tree oil's popularity may be due to such attitudes conveyed in the sayings "Fight fire with fire" or "Medicine has to taste bitter." This, however, is not in accord with the principles of holistic aromatherapy, which assume that a positive interaction between people and aromas has to take place. However, the liking or rejection of an aroma can also depend on the person's state of health. For example, a person who is ill generally finds the aroma of tea tree oil more pleasant than do people who are bursting with health.

The influence of tea tree oil on our mental and psychic states is not as strong as that of the alternative manuka and ravensara oils, which are described on pages 22 and 23. Nonetheless, tea tree oil is purported to support mental and psychic drives due to its balancing and nerve-strengthening effects.

Tea tree oil, however, doesn't have any soothing, comforting, or relaxing components. To create such effects, this classic Australian oil needs to be supplemented with lavender, geranium, chamomile, manuka, or ravensara.

Many people have to get used to the tangy herbal odor of tea tree oil.

Aromatherapy using tea tree oil can be carried out as follows:

- To steam the essential oil, put a few drops in a room humidifier, a diffuser, or an aroma lamp.
- Also, a few drops added to the bath water vaporize slowly and are taken in by the body through the nose and the skin.
- You can achieve the same effect when you mix tea tree oil with a massage oil and massage them into the skin.
- For inhalation during illnesses of the respiratory tract, use the methods described in the "Colds" section.

Animal Care

Even before tea tree oil's recent popularity, animal lovers were familiar with its use. It has mainly been used as a vermin repellant for horses and dogs. Tea tree oil is either applied neat or in a shampoo, and even some flea collars contain it. Of course, the essential oil and its dilutions are also suitable for disinfecting insect bites or stings, and the skin after the removal of ticks, as well as for healing abrasions or sore spots, caused by saddles, for example.

Cleaning and Disinfecting in the Home

Presently, laboratory tests are underway to find out the extent that tea tree oil can be used for disinfecting medical instruments. Preliminary results indicate that the concentrations needed are too high for such use, thereby making it probably unprofitable. However, using tea tree oil is basically a well-suited and environ-

Tea tree oil has long been in use in animal care.

mentally friendly way of cleaning the home, because our homes don't have to be completely sterile. Such a natural and harmless way of cleaning is especially desirable for household surfaces that come into contact with the human body, such as surfaces in the bathroom and kitchen counters. But before tea tree oil is used pervasively in the home, many tests will have to be conducted in order to recommend the proper mixtures and dilutions.

Applying Tea Tree Oil from Head to Toe

Tea Tree Oil in Cosmetic Products

Because the healing effects of tea tree oil have not yet been as thoroughly tested as the health authorities require for medicinal products, tea tree oil's primary area of importance is in external application, or body care. In addition to the pure essential oil and the hydrolate, there are quite a few excellent body-care products that contain tea tree oil. However, we caution against the overuse of tea tree oil, the reason being that a high dosage or constant use—this means using tea tree oil for just about everything—can lead to sensitization and to allergic reactions in people with sensitive skin.

Today, many manufacturers are careful to use other valuable natural substances in their products along with tea tree oil. A supplement to the German Cosmetic Regulations, instated in 1997, declares that from July 1998 onward, all the ingredients in cosmetics have to be stated on the packaging. There are similar regulations in other countries. This will help us be more knowledgeable consumers, especially when it comes to buying natural cosmetics.

Cosmetic—from the Greek word *kosmetikos,* meaning "skilled in adornment"—is more than the mere adornment of the body or the face, in particular. In fact, the tenets of beauty care are also concerned with health care, hygiene, and the prevention of illnesses.

Cleaning and Care of the Skin

Cleaning our bodies is one of the most important factors in keeping ourselves healthy. Just a few decades ago, however, this was by no means a matter of course. But increasing prosperity, hygiene, and education are why infections and illnesses that were caused by a lack of cleanliness (of the body, our food, and sanitary areas) are now rare in many places in the world.

Most people take a shower every day. However, some people have begun to overdo personal cleanliness. An increasing number of people suffer from dry, irritated skin due to too frequent bathing and showering, as well as to the use of aggressive soaps and cleansers. Cleaning the body aggressively withdraws the fat and moisturizing factors from the skin and destroys its natural protective film, which in turn leads to a higher susceptibility to fungal infections.

Excessive hygiene destroys natural skin protection.

This is how to look after your skin:

- Don't shower more than once a day, and only shower every two to three days if you have very dry skin. In between, wash only those parts of the body that are dirty or where body odor develops.
- Take a bath only once a week at the most. If you tend to have dry skin, you should take oil baths

(which do not clean very well, however).

- Always bathe and shower as quickly as possible, and never with water that is too hot.
- Do not use perfumed, aggressive soaps; slightly acidic cleansers are preferable.
- Don't forget to use a body creme or lotion after washing yourself.

Use of the Oil

Tea tree oil is also used in cleansing products, such as soaps and washing lotions, because of its antiseptic effect and good tolerance by the skin. As shown in the latest tests by Hammer, Carson, and Riley, the natural bacterial flora of the skin are definitely damaged less by small amounts of essential oil than by the germs causing illness. In addition, this disinfectant protects the cosmetic product it is in from being polluted by microbes so that, for the most part, it is possible to do without chemical preserving agents.

Tea tree oil is easy on the natural bacterial flora of the skin.

The fresh aroma as well as the neutral pH-value of tea tree oil are further properties that favor its use in skin and body care.

Oily skin, including the problem skin of adolescents, is very well suited to cleansing with tea tree oil. Especially those areas of the skin that have an excessive production of oil should be cleaned in the morning and in the evening using a wash lotion containing tea tree oil. Afterward, use a corresponding toning lotion and finally apply a face or body creme. Every now and then, you can disinfect the odd pimple with a cotton swab soaked in tea tree oil.

For Beautiful Hair and a Healthy Scalp

Clean and glossy, healthy, well-cared-for hair is obviously very attractive. But only a few men and women are truly happy with their hair, because they have problems with dandruff, itchy scalp, and hair that is either too oily or too dry.

Use of the Oil

It's possible to control most of these problems by using tea tree oil. Its effect on dandruff has already been described in the "Dandruff" section (page 54).

But tea tree oil can also be a great help if you have an itchy or an oily scalp because of its antibacterial and antiseptic properties.

Prevention of Body Odor

Without noticing it, an adult normally loses about a pint of water through the pores of the skin every day. High temperatures, stress, and strain, as well as excitement can increase this amount, which can, in extreme cases, come to more than 10 quarts a day. The body is able to regulate our temperature by perspiring, because the evaporation of the sweat on the skin has a cooling effect. However, an abnormally high secretion of sweat during periods of rest can be an indication of a hormone or a circulatory disorder, and you should see a doctor if such is the case.

Use of the Oil

Perspiration is usually colorless and odor-free. Only when it is broken down by bacteria on the surface of the skin does the odor arise. Deodorants can put a stop to this smell, which most people consider unpleasant. Tea tree oil is very well suited as an active agent in a deodorant, because the multiplication of the bacteria is stopped by such low concentrations of the oil that irritation of even the very sensitive areas of the skin, such as the armpits, is very rare. Furthermore, it absorbs and covers unpleasant smells with its fresh and "clean" aroma. You can find a suitable blend with other substances in ready-made tea tree oil deodorants, available at many health food stores and drugstores. Apply the deodorant to the cleaned armpits once a day. If necessary, this can be repeated after washing at night. Because underarm hair is a breeding ground for bacteria, it is recommended to shave the hair off if you tend to perspire a great deal.

Foot Care

Although the feet must endure a great deal of pressure and strain, they are ironically one of the most neglected parts of the body. In comparison, people generally take care and try to improve the appearance of their hands and fingernails. But many people seem to think that it is not worth taking the trouble to do the same with their feet and toenails. And why should they? Hardly anyone sees them, or so one might think.

Even if the feet are well hidden most of the time, they carry you throughout your entire life. Scientists once worked out that the average person in his or her lifetime walks as many miles as would equal circling the circumference of the Earth about four times!

Our feet deserve the best possible care.

However, instead of caring for the feet while they are performing this wearisome task, women often force their feet into shoes that are too tight and have heels that are too high and where they possibly even sweat more than necessary in nylon stockings.

Improper and badly fitting shoes can be the cause for later foot disorders, especially with children whose feet are very pliable and thus can be easily deformed.

Here are some tips for healthy feet:

Taking good care of the feet should be a matter of course from childhood onward.

- Shoes must fit well and be comfortable; the most suitable material is leather.
- Change your shoes regularly.
- Women should wear high heels in exceptional cases only, because they shift the weight in an unnatural way to the ball of the foot.
- A low heel of less than 2 inches is ideal for women.
- Be sure that your shoes have a well-formed lining made of natural materials (leather, felt, cork).
- Walk barefoot as often as possible (but not in wet areas).

You shouldn't wait for problems to occur to start taking care of your feet. To the contrary, regular pedicures prevent welts, corns, or ingrown toenails from developing at all, and they reduce perspiration of the feet. Those people who have to wear shoes made of synthetic materials during work (tennis shoes or rubber boots, for example) or people whose blood circulation is restricted due to a metabolic disorder, such as diabetes, have to pay extra attention to their feet to keep them healthy.

Use of the Oil

Start your foot-care program with a warm foot bath containing soap-free additives, because the alkaline content of soap would cause the skin to swell and soften too much. Special foot baths containing tea tree oil are recommended. One formula is adding to the bath water a solution of 20 drops of tea tree oil mixed into a little milk. The essential oil not only has antiseptic and deodorant effects, but it also improves the circulation and at the same time has a cooling result so that your feet feel a lot fresher afterward. If you suffer from cold feet, you can fill a second foot tub with cold water for alternate hot and cold foot baths.

Afterward, the feet have to be dried carefully, especially between the toes. You can now continue your foot-care program by smoothing down all callused areas of the feet with a pumice stone or a callus rasp (do not use a callus parer or a razor blade!). If you do this regularly, no true calluses will develop.

Always cut your toenails straight across—never in a curve—because curved toenails often grow into the tissue bordering the sides of the toenail.

After cutting your toenails, file down the sharp edges a bit. The cuticle may be pushed back carefully with a spatula made of wood or horn (never use force), but the cuticle must never be cut because this creates an opening at the nails and lets in fungi and bacteria.

Finally, you should renew the oil

and moisture of the feet with a foot-care creme or balm containing tea tree oil. This would be a good time for a foot massage, which relaxes muscles and tendons and thereby eases tension. A foot massage should always start from the tips of the toes and then proceed toward the body. A reflexology treatment would be ideal, and it need not necessarily be done by a professional because it isn't difficult to learn.

When you get home, don't stick your feet immediately back into shoes and socks. To round off your foot-care program, do some foot exercises—for example, grasp something with your toes, or roll your feet back and forth over a wooden roller—or merely walk around barefoot on the carpet; these are all very beneficial for your feet.

Of all the parts of the body, the sweat glands of the skin are closest together on the soles of the feet. If the feet get very warm—which happens easily when you are wearing socks or stockings made of synthetic materials or shoes not made of leather, because the air circulation around your feet is restricted—they start to sweat heavily. If the moisture cannot evaporate, it provides an ideal breeding ground for fungi, as already mentioned. Furthermore, if the sweat—which is 99 percent water, the rest being salts and organic substances—is broken down by bacteria, this results in that unpleasant foot odor.

Good hygiene is the most important factor in controlling sweaty feet and its consequences. By following the foot-care program described previously, you will already be well on your way toward preventing sweaty feet; especially the daily bathing and application of a creme with tea tree oil will make themselves felt, because they have a deodorant and cooling effect that lasts the entire day. Taking a foot bath twice a week, prepared as follows, has proven to be very effective in lessening perspiration of the feet.

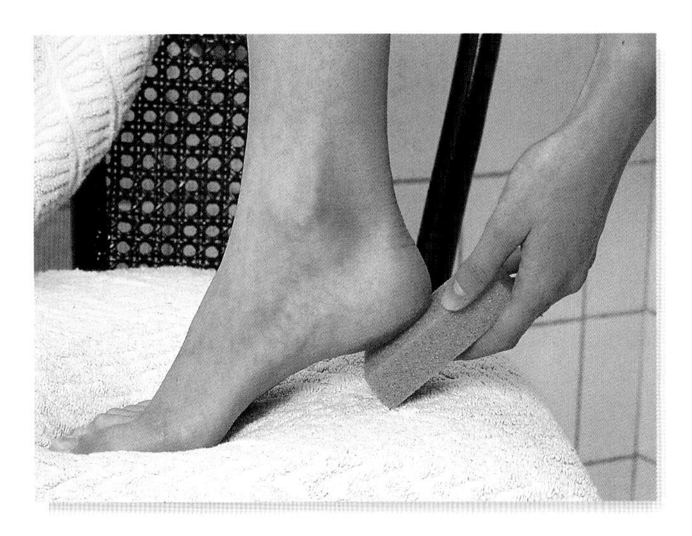

Remove calluses with a pumice stone and water from your foot bath.

Pour a pint of cold water over 3 tablespoons of oak bark, and heat it to boiling. After 20 minutes, strain the mixture. Pour it into your foot tub, and fill the rest of the tub with water from the faucet. Then, in a small bowl, mix 15 drops of tea tree oil, 15 drops of balm oil (you can also make a tea with the dried balm leaves), and a little milk, and add the solution to the water. Bathe the feet in this for about 10 minutes.

Along with sweaty feet, callused skin is one of the most frequent disorders of the feet that you can control by diligent foot care. The skin of the feet can thicken and become hard when subjected to strong pressure. And because there is no sebum—necessary to keep the skin soft—in the hard, thickened layer of skin, it can become crumbly and split open in some places. Such cracks are openings for bacteria and fungi as well as for the viruses that cause warts. So, you can see that excessively hard, thickened skin is not just a cosmetic problem.

With corns, the callus grows like a thorn inward.

Corns are a special case, whereby constant pressure on one spot causes the callus to grow like a thorn down into the skin tissue, which is rich in nerves, thus producing pain. For both calluses and corns, the pressure causing the problem has to be removed and the hard, thickened skin has to be taken off with chemical or natural preparations, or a corn-cutter, in the case of corns.

Besides being used in foot baths and cremes, tea tree oil can be applied in a concentrated form to treat corns. What is recommended is to stick a Band-Aid soaked in pure tea tree oil onto the corn and change it daily. (The sticky part of the Band-Aid should not come into contact with the oil, because it would then dissolve.) In this way, the corn is softened gently. After a week of this treatment, soak your feet for about 10 minutes in soap suds and then carefully remove the hardened skin with a corn remover. This procedure has to be repeated until the entire corn down to the root has been removed.

Skin Care After Sunbathing

We become aware of the need for a sun blocker, as well as just how hard it is to relieve sunburn, in the evening after sunbathing during a scorching day, when the skin has tightened up, it begins to itch, and it longs for moisture and cooling. Moisturizing lotions that contain cooling and anti-inflammatory substances are important now. On the other hand, oily cremes and suntan oils are not recommended, because they quickly lead to heat building up in the skin.

Use of the Oil

In addition to chamomile and aloe vera, tea tree oil is recommended for sunburns, because it has a cooling effect, relieves the itching and the pain, and soothes the skin. Suitable for the care of sun-damaged skin are the commercial body lotions or après-sun preparations containing chamomile, aloe vera, panthenol, and vitamin E, to which a few drops of tea tree oil have been added.

Dental Hygiene

Sparkling, white teeth in slightly pink gums—unfortunately, few of us have teeth as dazzling as those in the toothpaste advertisements. To the contrary, many people already had tooth decay in childhood, and nearly every adult suffers to some extent from receding of the gums. As much as many of us would like to blame tooth

decay and gum problems on poor genetics, the truth is that they are more often due to faulty dental hygiene.

As with the skin, bacteria are natural and normal in the mouth; this is what is known as the flora of the mouth. However, if the bacteria multiply too rapidly—and this can happen quite quickly under certain circumstances—they can cause damage to the teeth and the gums. They adhere to the teeth after meals, and, together with the food particles between the teeth and the dead cells of the mouth flora, they form plaque, which is primarily a soft and yellowish film. In this film, the bacteria then produce acids from the food sugars, which, if not neutralized by saliva, can damage the tooth enamel by withdrawing minerals, such as calcium and phosphate. If the enamel is not remineralized by fluorides, after a while holes will develop in the teeth.

Plaque has to be removed as quickly as possible, because only after a few days it turns into tartar, which is difficult to dislodge. Although not harmful as such, the rough surface of the tartar makes it easier for new plaque to be deposited. And this process does not stop at the gums, which form a protective covering around the neck of the tooth and the upper parts of the root of the tooth. If the incrustation constantly irritates the gums, they recede. The neck of the tooth is no longer protected from an attack, and the tooth loosens. Even healthy teeth can be lost this way.

Proper oral hygiene is an important preventive measure, and necessary for keeping your own teeth for as long a time as possible. For proper oral hygiene, follow these important guidelines:

- Ask your dentist to remove old, hard incrustations of tartar.
- Toothbrush: Soft to medium, rounded bristles (no natural fibers); your toothbrush should be replaced every two months.
- Cleaning technique: Small, circular, and shaking movements from the gums to the crown of the tooth; dentists recommend using electric toothbrushes to remove plaque—and, by the way, it is better to brush your teeth twice a day thoroughly than to do so sloppily after each meal.
- Toothpaste: Should contain fluorides (aminofluoride) and antiseptic substances (tea tree oil, for example) as well as the cleansing

Plaque is formed from bacteria, food particles, and dead mucous membrane cells.

substances (calcium carbonate or calcium phosphate); do not use abrasive substances (toothpaste for smokers, tooth bleaches)!

- Gaps between the teeth: Clean using dental floss (for beginners, it is easier to use waxed floss) and interdental brushes (from the drugstore).
- Mouthwashes: Hinder the growth of bacteria and freshen the breath; they contain fluorides and antiseptic substances, and can be used to rinse out your mouth in between meals.

To prevent tooth decay, your toothpaste must contain fluoride.

Use of the Oil

Tea tree oil is very helpful in oral hygiene because of its antiseptic effect. If used regularly, it can prevent cavities and gum problems. Furthermore, its freshness combats bad breath and a sour taste in the mouth.

For the daily cleaning of your teeth, just add a few drops of tea tree oil to a toothpaste containing aminofluorides. After brushing your teeth, and in between (if possible after every meal), rinse your mouth and gargle with half a glass of warm water to which 1 to 2 drops of tea tree oil have been added.

Toothpaste containing tea tree oil is now available commercially. However, if these products don't contain fluorides, they are suitable only as a support to your dental hygiene.

Many Australian dentists, and more and more of their colleagues in other countries, not only recommend tea tree oil to their patients as a preventive for various tooth ailments, but they are also using it in treating their patients, for example, after the surgical extraction of a tooth. They apply a strip of gauze soaked in tea tree oil to the injured spot to disinfect the wound and ease the pain.

Appendix

Glossary

AIDS: Acquired immunodeficiency syndrome, caused by infection with HIV.

Antimycotic: Effective against fungi.

Antiviral: Effective against viruses.

Aphthae: Sharply defined, small, and painful festering blisters on the mucous membranes, mainly in the mouth.

Chemoprophylaxis: The prevention of infectious diseases (malaria, for example) through the use of chemical agents.

Comedones: Blackheads.

Cortisone: Hormone of the adrenal cortex; it is used in the treatment of various inflammatory, allergic diseases.

DDT: Dichlor + diphenyl + trichlor, an insecticide that was used successfully in the fight against malaria, but that tends to accumulate in ecosystems and has toxic effects; since 1972, its use has been banned in Germany.

Diabetes mellitus: A variable disorder of carbohydrate metabolism; an inadequate production of insulin by the pancreas causes excessive amounts of sugar in the blood.

Diffuser: A fan-driven device for dispersing aromatic oils.

Emulsion: The dispersion of one liquid within another that usually cannot be mixed; mostly finely dispersed oil in water, or vice versa.

Fungicide: An agent that kills fungi or stops their growth.

Galenical: A medicine prepared from constituents of plants.

Gauze: A loosely woven surgical dressing often soaked with medicinal preparations.

Histamine: A crystalline compound found in plant and animal tissues; it is released from the mast cells of the tissues during allergic reactions, and typical symptoms are reddening of the skin, itching, and the development of rashes.

Immunosuppressants: Suppress the natural immune responses of the body; they are used, for example, when transplanting organs so that the new organ will not be rejected by the body.

Intestinal flora: Natural bacteria cultures in the intestines.

Irritation: A sore, irritated, or inflamed condition of a bodily part—the skin, for example.

Lindane: An insecticide, small amounts of which are poisonous to people; also a medication used to get rid of parasites of the skin, such as scabies and lice.

Local anesthetic: An anesthetic that acts only on and around the point where it is applied.

Machete: A large, heavy bush knife.

Mycosis: An infection with, or a disease caused by, a fungus.

Oleic gauze: A wound dressing soaked with oil or a liquid similar to oil that keeps the dressing from sticking to the wound.

Oral: Through the mouth.

Papules: Small, solid knots in the skin.

Pathogen: A specific cause of a disease.

Placebo: An innocuous substance used especially in controlled tests studying the efficacy of another substance, such as a medicine.

Pustule: Small elevation of the skin containing pus and having an infected base.

Receptors: Specific connecting links in the body for transmitting stimuli.

Remineralization: Resupplying the body with lost mineral salts.

Systematic therapy: A form of therapy whereby the patient takes or injects a medical substance, which then spreads throughout the entire body and thereby also reaches the seat of the illness.

Toxic: Relating to, or caused by, a poison.

Trichomoniasis: A sexually transmitted disease occurring especially as vaginitis and caused by a trichomonad that may also invade the male urethra and bladder.

Virus: The causative agent of an infectious disease, capable of growth and multiplying only in living cells.

Checklist: Staying Healthy When Traveling

What You Should Take from Home

Minimum Set:

- One small bottle of tea tree oil
- One bottle (100 ml) of 70 percent isopropyl alcohol
- One empty 20-ml bottle for mixing
- Band-Aids

Additional Set:

- Lip balm containing tea tree oil
- Body oil and after-sun lotion
- Insect repellent (homemade, see page 64)
- Creme containing tea tree oil
- Shampoo containing tea tree oil

Complaints	Use
Blisters	Don't open blisters! Dab them with neat tea tree oil, and let them dry. Then cover them with a gelatinous, hypoallergenic bandage.
Disinfecting bites	Put some pure tea tree oil or a 1:10 solution cut with 70 percent alcohol on the wound.
Disinfecting toilets or sanitary facilities	Put some pure tea tree oil or a solution of 1:10 with 70 percent alcohol onto a paper tissue, and wipe with this.
Foot disorders	For a foot bath, mix 20 drops of tea tree oil with a little milk and add to warm water. Bathe your feet for about 10 to 15 minutes, and dry them thoroughly afterward.
Athlete's foot	Foot bath, as above. In addition, rub the feet, especially in between the toes, with a mixture of tea tree oil and the alcohol in a proportion of 4:6 parts.
Corns	Put 1 drop of tea tree oil on the corn, and cover with an adhesive bandage. Do not let the oil run onto the adhesive, because this could irritate the skin.
Inflammations of the throat	Add 2 to 3 drops of tea tree oil to half a glass of warm water. Gargle twice a day.
Inflammation of the skin	Dab the affected areas with a 4:6 mixture of tea tree oil and the alcohol.
Insect repellent	Add the alcohol to 2 ml of tea tree oil to make up 20 ml. Apply this to the skin every 4 hours. Or use homemade insect repellent (page 64).

Complaints	Use
Insect stings	Dab with neat tea tree oil.
Itching	Apply a creme or lotion containing tea tree oil (1 drop of the lotion in the cup of your hand).
Lice	Add 30 drops of tea tree oil and, if required, 20 drops of lavender oil to 100 ml of shampoo. Shampoo your hair, leave the shampoo in for 30 minutes, and rinse. Then remove the eggs with a lice comb. Repeat every 3 to 4 days.
Lip blisters	Dab with pure tea tree oil as soon as you feel a tingling sensation. In addition, use a lip balm containing tea tree oil.
Aching muscles	Add 2 to 3 drops of tea tree oil to some body lotion in your cupped hand, and massage the muscles with this mixture.
Inflammation of the mucous membranes of the mouth	Add 2 to 3 drops of tea tree oil to half a glass of warm water, and rinse your mouth.
Colds	Add a few drops of tea tree oil to some water; soak towels in this solution, and hang them up in your room. Or add 3 drops of tea tree oil to a cup of water, and leave it to evaporate on a tea-warmer. Or fill a bowl with a quart of boiling water, add 2 to 5 drops of tea tree oil, and inhale with a towel over your head.
Sunburn	Gently apply a creme or after-sun lotion containing tea tree oil (1 drop in a handful of lotion) to the skin.
Vaginal infections	Dilute 70 percent alcohol at 1:4 with (boiled) water. Mix 6 ml of this with 17 percent alcohol and 4 ml of tea tree oil, and add it to a hip bath. Or soak a tampon in a 1:10 solution (boiled water) of this tea tree oil mixture, and insert it into the vagina.
Burns (minor)	Run cold water over the skin for 10 to 20 minutes. Then dab with neat tea tree oil.
Inflammation of the gums	Add 2 to 3 drops of tea tree oil to half a glass of warm water, and rinse out your mouth.
Toothache	Dab neat tea tree oil on the affected tooth and the surrounding gums.

Index

A

Aborigines, 10, 25
abrasions, 53–54
abscesses, 43
acne, 32–33, 43–47
allergens, 68, 69
allergic reactions, 37, 38
animal care, 76
antibiotics, 35
antifungal diet, 65–66
antifungals, 27, 32
antiseptics, 25, 29–30
aromatherapy, 74–76
athlete's foot, 47–49, 90
ATTIA (Australian Tea Tree Industry Association), 19, 39
Australian Aborigines, 10, 25
"Australian Gold," 11
Australian Standard, 18
Australian Tea Tree Industry Association (ATTIA), 19, 39

B

balm mint, 66
blackheads, 44
black tea, 19
blisters, 66–67, 90, 91
body odor, preventing, 81
boils, 30
borage, 69, 70
burns, minor, 49–50, 91

C

cabbage palm *(Coryline australis),* 19
cajeput oil, 20–21
calluses/corns, 84, 90
Camellia sinensis, 19
Candida albicans, 27, 30–31, 57
chicken pox, 50
cineole 1.8 (eukalyptol), 18, 36–37, 52
cleaning
  hair/scalp, 80–81
  household, 76–77
  skin. *See* skin care
climate, for tea tree growth, 13–14
clinical studies, 29–33
cold, common, 51–52, 90
comedones, 44
components
  of essential oils, 16
  of fatty oils, 16
  of tea tree oil, 17
contact allergy, 68, 69
Cook, Captain James, 10–11
corns/calluses, 84, 90
cosmetic products
  comedogenic ingredients, 44
  tea tree oil in, 79–86
coughs, 52–53
cuts, 53–54

D

dandruff, 54–55
dental hygiene, 84–86
dexpanthenol, 61
diaper rash, 62
diet, antifungal, 65–66
disinfectant properties
  of essential oils, 10, 25–26
  of tea tree oil, 25–26, 29, 76–77, 90
distillation process, 16

E

emulsion, tea tree oil, 58
Escherichia coli, 26
essential oils (etheral oils)
  cajeput, 20–21
  disinfectant properties, 10
  general properties, 16–17
  manuka, 22–23
  naiouli, 21–22
  ravensara, 23
  tea tree. *See* Tea tree oil
eukalyptol (1.8 cineole), 18, 36–37, 52

external application, 36
extraction of oil, by distillation, 16

F

fatty oils, 16
first-aid travel kit, 73–74
floss, dental, 86
food allergens, 68
foot
    athlete's, 47–49, 90
    calluses/corns, 84, 90
    care, 81–84
    problems, 31–32, 90
    sweaty, 83–84
fungal infections
    comparative efficacy, 27
    foot, 32
    intestinal, 65–66
    nail, 59–60
    vaginal, 55–59
furuncles, 43

G

gangrene, 61
German Book of Medicines, 39
German Cosmetic Regulations, 79
"greenhouse effect," 56
gum inflammation, 61–62, 91

H

hair cleaning/care, 80–81
harvesting, of tea tree oil, 14–15
healing, 29–30
herpes virus, 66
historical aspects, 10–12
hoarseness, 71
hydrolate, 16, 42

I

inflammation
    of gums, 61–62, 91
    of mucous membranes in mouth, 91
    of nail wall, 62
    of skin, 62–63, 90
    of throat, 90
insect bites/stings, 63, 90, 91
insecticides, 15
insect repellent, 63–65, 90
internal usage, 38, 63, 66
intestinal fungi, 65–66
itching, 91

J

joint pain, 70–71

L

laboratory tests, 25–29
lactic acid bacteria, 59
Leptospermum scoparium, 22–23
lice, 91
lindane, 15
lip blisters, 66–67, 91
local anesthetic, 69, 70
Lyme disease, 63–64

M

manuka oil, 22–23
massage, foot, 83
Melaleuca alternifolia, 13, 19, 39
Melaleuca dissitiflora, 13, 19, 39
Melaleuca leucadendron, 20–21
Melaleuca linariifolia, 13, 19, 39
Melaleuca viridiflora, 21–22
melissa, 66
"Miracle of the Island Continent," 12
mosquito bites, 63
mouth mucous membranes, inflammation of, 91
mouthwash, 86
muscle pain/aching, 70–71, 91

N

nail fungus, 59–60
nail wall inflammation, 62
natural allergens, 68
neurodermatitis, 67–70
niaouli oil, 21–22

O

oil baths, 68–69
oral hygiene, 85–86

P

pain
    muscles/joint, 70–71, 91
    toothache, 72, 91
panaceas, 25

paperbark tree, 14
paracymene, 37
pathogens, susceptible to tea tree oil, 26
Penfold, Dr. Arthur R., 11, 25
perishability, 39
perspiration, 81
phenol, 11, 25
phytotherapy, 11–12
pimples. *See* acne
Ping-Pong effect, 58
Pityrosporum ovale, 55
plaque, dental, 85
Propioni acnes, 44
Propionibacterim acnes, 26
Pseudomonas aeruginosa, 26

Q
quality standard, for tea tree oil, 18–19

R
Ravensara aromatica, 23
ravensara oil, 23
rosacea, 71
"rosy belt," 50
Rudbeckia hirta, 52

S
safety, 39
scalp cleaning/care, 80–81
shingles, 50
skin care
  for acne, 46–47
  after sunbathing, 84
  daily, 79–80
skin inflammation, 62–63, 90
solanum, 69, 70
solutions, tea tree oil, 41–42
sore throat, 71, 90
Staphylococcus aureus, 26
storage, 39
Streptococcus faecalis, 26
Streptococcus pyogenes, 26
stress, 68
sunbathing, skin care after, 84
sunburn, 91

T
tea tree
  botanical aspects, 13–15
  name, 10, 11
  relatives, 19–23
tea tree oil
  components, 17
  future of, 12–13
  location, in tree, 14
  mechanism of action, 35
  mixing with water, 41–42
  neat or undiluted, 41
  odor, 17
  properties, 27–28
  quality standard, 18–19
  success story of, 10–13
  usages. *See* specific disorders
tea tree plantations, 14–15
terpinen-4-ol, 18, 35
throat inflammation, 90
ticks, 63–64
titrea, 19
toenail care, 82
toilets, disinfecting, 90
tolerance, 36–37
toothache, 72, 91
toothbrush, 85
toothpaste, 85–86
trichomonads, 30

V
vaginal infections, 30–31, 55–59, 91
Varicella zoster, 50
viral infections, 27–28, 50

W
warts, 72–73

Y
yogurt, 59

About the Author

Susanne Poth is a pharmacist. She has become well known through many of her articles in professional journals as well as consumer magazines, and as an author of several books on medicinal plants.

Translated from the German by Elfie Homann and Range Cloyd

Edited by Laurel Ornitz